Alistair Tough

THE SHAPING OF THE MEDICAL PROFESSION

The Shaping of the Medical Profession

The History of the Royal College of Physicians and Surgeons of Glasgow, 1858–1999

Andrew Hull and Johanna Geyer-Kordesch

THE HAMBLEDON PRESS

London and Rio Grande

Published by The Hambledon Press, 1999

102 Gloucester Avenue, London NW1 8HX (USA)
Po Box 162, Rio Grande, Ohio 45674 (USA)

ISBN 1 85285 187 2

A description of this book is available from the
British Library and from the Library of Congress

Typeset in Minion by Carnegie Publishing, Lancaster
Printed and bound in the UK on acid-free paper
by Cambridge University Press

Contents

Illustrations

17 Sir Donald Campbell, PRCPSG, welcoming Diana Princess of Wales to the College on the occasion of the opening of the Princess of Wales Conference Suite, 8 June 1993. *(RCPSG)*

Abbreviations

ad eundem	A method of admission to the Fellowship by which the candidate's previous medical qualifications are recognised as of equal status.
AGM Billet	The Billet announcing the Annual General Meeting of the College, which used to be in November of each year, but is now held in December. All billets are held in RCPSG 1/1/5.
ASGBI	Association of Surgeons of Great Britain and Ireland.
BDS	Bachelor of Dental Surgery.
BMJ	*British Medical Journal.*
CCPME (GB)	Central Committee on Postgraduate Medical Education of Great Britain (from 1967).
CCST	Certificate of Completion of Specialist Training.
CVCP	Committee of [University] Vice-Chancellors and Principals.
DCH	Diploma in Child Health (from 1957).
DHS	Department of Health for Scotland.
DQ	Double Qualification.
EMJ	*Edinburgh Medical Journal.*
EMS	(WW2) Emergency Medical Service.
FDS	Fellowship in Dental Surgery (from 1967).
FPSG	Faculty of Physicians and Surgeons of Glasgow (the title from the beginning of this volume to 1909).
GDC	General Dental Council.
GMC	General Medical Council.

GMJ	*Glasgow Medical Journal.*
GPGMA	Glasgow Postgraduate Medical Association. Operated 1914–1944.
GPGMB	Glasgow Postgraduate Medical Board (an autonomous Board with the same bias in representation as the GPMEC – but on both 'University' members might also be key Faculty men). Operated from 1960–1970).
GPGMEC	Glasgow Postgraduate Medical Education Committee (a Committee of the University Medical Faculty on which the Faculty had four representatives to the University's eight). Operated from 1944–1960).
GPT	General Professional Training.
GRI	Glasgow Royal Infirmary.
GUABRC	Glasgow University Archives and Business Records Centre.
GWI	Glasgow Western Infirmary.
HDD	Higher Dental Diploma (from 1920).
ICSF	Intercollegiate Specialty Fellowship.
JCC	Joint Consultants Committee.
JCHMT	Joint Committee on Higher Medical Training (from 1970).
JCHST	Joint Committee on Higher Surgical Training (from 1969).
LDS	Licence in Dental Surgery (from 1878).
MAC	Medical Advisory Committee of College Council. (from 1962).
MRC	Medical Research Council.
NHI	National Health Insurance (from 1911 Act).
NHS	National Health Service.
NPHT	Nuffield Provincial Hospitals Trust.
RCPL	Royal College of Physicians of London.
RCGP	Royal College of General Practitioners (founded 1952).
RCSEng	Royal College of Surgeons of England.

RCOG	Royal College of Obstetricians and Gynaecologists (founded 1929).
RCPEd	Royal College of Physicians of Edinburgh.
RCPI	Royal College of Physicians of Ireland.
RCSEd	Royal College of Surgeons of Edinburgh.
RCSI	Royal College of Surgeons in Ireland.
RCPSG	Royal College of Physicians and Surgeons of Glasgow (the post-1962 title). In footnotes, when accompanied by a reference number (as in RCPSG 1/1/1/17) this refers to the Royal College's Archive Collection.
RFPSG	Royal Faculty of Physicians and Surgeons of Glasgow (the title from 1909–1962).
RHB	Regional Hospital Board.
RHSC	Royal Hospital for Sick Children.
RMCSG	Royal Medico-Chirurgical Society of Glasgow.
RMS	Royal Medical Society.
PRCPSG	President of the RCPSG.
SBC	Scottish Branch Council (of the GMC).
SAC	Surgical Advisory Committee of College Council (from 1962).
SCPGME	Scottish Council for Postgraduate Medical Education (from 1970).
SHHD	Scottish Home and Health Department.
SJC	Standing Joint Committee of the three Scottish Royal Colleges.
SMO	Specialist Medical Order (*Statutory Instruments*, 1995 no. 3208, 'Medical Profession: The European Specialist Medical Qualifications Order 1995'. The entire order came fully into effect from 1 January 1997).
SPMA	Scottish Postgraduate Medical Association (from 1962).
ST	Specialist Training (also called Further Professional Training).

STV   Scottish Television (now part of Scottish Media Group).

TQ   Triple Qualification.

VI   Victoria Infirmary, Glasgow.

WRCPGME   Western Regional Committee for Postgraduate Medical Education. Operated 1970–1974.

WRHB   Western Regional Hospital Board.

WSCPGME   The WRCPGME was renamed the West of Scotland Committee for Postgraduate Medical Education in 1974, reflecting the fact that it covered all western Health Boards under the 1974 NHS reorganisation.

Introduction

This volume of the history of the Royal College of Physicians and Surgeons of Glasgow takes up the story from where Johanna-Geyer-Kordesch and Fiona Macdonald's first volume, *Physicians and Surgeons in Glasgow, 1599–1858*, left the then Faculty in 1858: having just achieved a seat at the top-table of national licensing bodies. It covers a period in which the nature of medical knowledge and the organisation of medical practice changed dramatically. It deals with a period in which the ancient medical corporations had increasingly to fight to retain professional self-regulation against the encroachments of the regulatory state. In this context the Glasgow body initially lost the official role in granting first registrable qualifications in medicine it gained in 1858. Then, as the growing corpus of medical knowledge became increasingly scientific and specialised, the Faculty (subsequently the College) gained, and retained, a new role as both an examining and teaching body in the postgraduate education of hospital doctors. The Glasgow body adjusted its role from granting some of the first to granting some of the last qualifications of the medical career.

While this volume will chart these changes, in both the local and the national contexts of changing medical knowledge, education and practice, it is primarily the history of an institution. The focus is necessarily squarely on the College itself, and how it achieved its key role in postgraduate education. The book will serve as an essential complement to histories of the University Medical Faculty and of the major Glasgow hospitals.

In 1988, in a thinly disguised parody of the twentieth-century history of the Royal College of Physicians and Surgeons of Glasgow a certain 'Timothy K. Coniston, FRCP (Chow.)' described the sad narrative of the Royal College of Surgeons and Physicians of Chowbent:

> Early in the present century it lapsed into a series of financial crises. It was almost destroyed by two world wars and the sequel. By this I mean the national reorganisation of the status of the old medical incorporations ... *vis-à-vis* the medical schools of the universities, all enjoying high incomes provided by the Exchequer. Briefly, the Institute's principal source of income, candidates' 'conjoint fees', dried up almost completely. There were no reserve funds of any importance, for the Institute had been content to live from hand-to-mouth ... this licensing body

became an object of derision among the medical practitioners in the Pennines, men who should certainly have been ashamed of their disloyalty ...

He concluded that:

> You can fiddle about with high-falutin titles, you can quote your fascinating historical background, and luxuriate in your splendid old library, but if you don't have enough money to pay your bills and provide a meeting place that compares with a very good hotel, you are wasting your time, you are a non-entity.[1]

This whimsical piece, written from the relative financial security of the late eighties, reminds us of a fundamental truth about institutions: they need sufficient income to survive. In the article, written under a *nom-de-plume* by Stanley Alstead,[2] one of the pivotal figures in the modern history of the RCPSG, it is suggested that it was the introduction of a compulsory annual subscription from 1 October 1956 which saved the then Faculty from disaster. Certainly this move secured the immediate survival of the Faculty; but without a clear professional role in the linked definition and organisation of medical knowledge and practice in Britain, in the mid to late twentieth centuries, it is doubtful whether even local medical men would have continued to become Fellows of what would have been merely an elite local professional medical club. A vibrant professional body must have a relevant role.

By the mid fifties, the Faculty had, however, already begun to reinvent itself as the local postgraduate teaching and examining centre offering higher qualifications for would-be consultants in the new NHS of equal quality and status to any other national qualification. Putting the Faculty on a secure financial footing was part of Alstead's plan, but it went hand-in-hand with a new role at the centre of postgraduate teaching and examining. Alstead, as Visitor and President, did not act alone, of course. As Tom Gibson has reminded us:

> The fact that during the Presidency of X, item Y occurred or was brought to fruition does not necessarily mean that President X was the prime mover or the effective agent of Y. Previous Presidents, other office-bearers, members of Council and individual Fellows and Members of the College may have done as much as, if not more than, the President of the time.[3]

---

[1] 'What's in a Name: A Dialogue on Colleges, Faculties and Institutes', Timothy K. Coniston, FRCP (Chow.), *College Bulletin*, May 1988, vol. 17, no. 3.

[2] His other literary alias was Rivington Pike. In his later years Alstead took up residence in the Lake District, hence both these pen-names and the location of the mythical college. I am grateful to Ian D. Melville, FRCP Glasg., Editor of the *College Bulletin*, for this information.

[3] Tom Gibson, *The Royal College of Physicians and Surgeons of Glasgow: A Short History Based on the Portraits and Other Memorabilia* (Edinburgh, 1983), p. 253.

The first chapter of this volume describes how the Glasgow Faculty struggled to get its qualifications generally accepted in the period *after* the Medical Act of 1858. Whilst the Act recognised the Faculty as awarding a registrable qualification, this state sanction was not enough. While the state was concerned with standards, and not breadth of curriculum, the market for medical expertise had already moved on and was demanding double qualifications in medicine and surgery. A watershed in the evolution of the modern profession, the Act was nevertheless loosely worded and difficult to apply. Like most legislation, its exact meaning had to be worked out in practice, in Scotland, in England – and then in Britain. The Faculty joined first with the Edinburgh Physicians to institute the Double Qualification in 1859, and then with the Edinburgh Physicians and Surgeons to award the Triple Qualification in 1884. These moves were both attempts to placate the medical marketplace and also to preempt further state control of the profession by voluntary reform. John Simon, Medical Officer to the Privy Council, supported the Army and the Poor Law Board in wanting a double qualification. But his aim was a single portal of entry to the medical profession: one universal state-monitored examination for all doctors. The Faculty and the other corporations argued for professional self-regulation in a free market.

In spite of these moves, the corporations struggled to maintain the popularity of their qualifications against those of the universities. The Carnegie Trust's grants to university students played a role in this from 1901. Extramural students did not qualify for these grants and therefore the popularity of these schools, whose students in Glasgow largely took the Faculty's TQ, waned. But changes in the nature of medical knowledge were also important in making university degrees more popular. As medicine became more scientific, it was increasingly only the university sector that could afford the large laboratories needed for mass teaching. The second chapter describes the decline of the Royal Faculty (its title from 1909) as the provider of first registrable qualifications in medicine, in a period when the university degree rose in popularity up to 1930.

The remaining chapters describe how the Royal Faculty (from 1962 Royal College) successfully moved into a new role in the postgraduate teaching and examining of hospital doctors. This is the heart of the book. By the 1940s, the British state had become committed to a particular ideology of medicine, a particular view of what constituted modern medical knowledge and practice. Progressive, up-to-date medicine was perceived to be specialised, scientific and hospital-based. It was ideally practised by highly-trained consultants, working in specialised units, who combined clinical research using laboratory methods with their patient care. In short, the state had taken up an academic conception of medicine and had adopted as an ideal the types of university-hospital relationships found in America and Germany.

Academic ways of thinking and of working were recommended in the influential Goodenough Report of 1944, which argued that the local university should be the centre of regional medical education. On the report's recommendation, the universities absorbed the extra-mural schools of medicine in 1948, and the university degree became, to all intents and purposes, that single portal of first entry to the profession that John Simon had hankered after. Furthermore, university medical schools began to strengthen their links with the hospitals by establishing full-time clinical chairs. This was an attempt to control the nature and scope of the clinical teaching of future generations of doctors and thereby to infuse them with the new academic agenda in medicine.

The hospital- and consultant-based NHS system of health care provision, which was regionally organised around central universities, together with the new educational responsibilities of the regional universities, provided for the first time a large number of full-time posts in the hospitals. This greatly increased the demand for highly-trained hospital doctors, and therefore for formalised postgraduate training and examinations. Hospital doctors now took a university degree and then built up clinical experience in a succession of hospital posts, while developing a special interest in a particular area of medicine. But there was little structure to the clinical experience which constituted this continuing period of medical education and training after initial registration. There was no theoretical, systematic lecturing at all. Yet this postgraduate period was increasingly important as medicine became more scientific and specialised, as well as government funded. From the 1940s, there was thus an increasingly pressing need for the central organisation of postgraduate education and a standardised system of examination. Just what a consultant was, what he or she should be expected to *know*, needed to be established. And, as the medical corporations had learned from 1858, if they didn't get there first, the state would.

How were these national trends reflected in developments in Glasgow? Malcolm Nicolson has argued that Scottish, and especially Glaswegian, medicine was long characterised by the epistemological hegemony of the clinic over the laboratory. Even when scientific departments (such as Physiology, Bacteriology, Microbiology) emerged, their work was perceived by their own staff and by the clinical elite as subservient to the clinic's priorities. They were there to test specimens and theories derived from clinical practice. Clinical practice, and not the laboratory, was where medical knowledge was advanced. As Nicolson wrote in 1993:

Overall ... the Scottish Universities were relatively slow to exploit the full potential of the new disciplines of scientific medicine ... This was, in part, due to the distinctive character of the Scottish medical schools. Strong commitments to medical education and clinical treatment produced an ethos which discouraged physiologists and biochemists from pursuing research projects which did not carry an immediate clinical implication ... It was to be some time before the Scottish medical faculties fully resolved the problem of accommodating laboratory science within institutional structures in which clinical priorities were necessarily dominant.[4]

In Glasgow, for much of the period covered by this volume, a medical culture existed similar to that identified by Christopher Lawrence among the elite London physicians of the late nineteenth and early twentieth centuries.[5] The clinical elite which controlled the hospitals, and made up the Faculty, also controlled what counted as medical knowledge. Some of the more progressive may have utilised modern scientific techniques in diagnosis, and even therapy, but they believed that the final arbiter in any case was clinical acumen built up through long clinical experience. Clinical experience was the basis of all medical knowledge, and thus the cornerstone of medical education. The social corollary of this was that most hospital doctors held part-time appointments, earned their money in private practice, and did little or no research, since professional advancement was not based on research but on clinical experience and social connections. In short patterns of working reinforced patterns of thinking and vice-versa.

Chapter 3 discusses the watershed in methods of working and thinking for Glasgow medicine which came from 1936 with the arrival of Sir Hector Hetherington as new University Principal and Vice-Chancellor. Hetherington brought the new academic medicine to Glasgow (in advance of both Goodenough and the NHS Acts) with a series of key full-time appointments to clinical Chairs in the major Glasgow hospitals. The men he appointed all had strong scientific interests and ways of working, and they built up around them research schools based on these methods. This was the beginning of setting in place Nicolson's 'institutional structures' for accommodating laboratory methods into the clinical side of university medicine.

The new academic priorities were, of course, not initially popular with the Glasgow clinical elite, as represented by the Faculty. The Faculty was, however,

[4] Malcolm Nicolson, 'Medicine', in P. H. Scott (ed.), *Scotland: A Concise Cultural History* (Edinburgh, 1993), pp. 327–42. Quotation from p. 340. See also Stephen Jacyna, 'The Laboratory and the Clinic: The Impact of Pathology on Surgical Diagnosis in the Glasgow Western Infirmary, 1875–1910', *Bulletin for the History of Medicine*, 62 (1988), pp. 384–406.

[5] Christopher Lawrence, 'Incommunicable Knowledge: Science, Technology and the Clinical Art in Britain, 1850–1914', *Journal of Contemporary History*, 20 (1985), pp. 503–20.

an institution increasingly without a meaningful role in the changing world of medical knowledge and practice. Hetherington offered the Royal Faculty a tacit bargain: cooperation with the academicisation of Glasgow medicine would result in a new role for the Faculty. The University would use the Faculty as its postgraduate arm: it would be helped (including financially) to establish itself as a postgraduate teaching and examining body to provide high-quality local training for hospital doctors. Even if, as increasingly happened, Fellows and Members of the Glasgow Faculty were foreign doctors, it would nevertheless provide a local intellectual focal point in the postgraduate sphere; and, in so doing, it would become an integral part of the academicisation (or modernisation) of Glasgow medicine.

Relations with the University, which, as *Physicians and Surgeons in Glasgow* shows, were strained when the two institutions were competing for students, now became symbiotic. This was helped by the fact that most key Faculty/College personnel were also now key University players. The 1950s saw the Faculty establishing its postgraduate teaching programme which prepared candidates for the Fellowship, now once again in demand as a passport to the road to consultant status. The 1960s saw the new Royal College of Physicians and Surgeons of Glasgow working with the other Royal Colleges to standardise postgraduate qualifications to fight off a further challenge from the Todd Committee's Report. The Todd Report provided a much-needed and widely-supported template for the organisation of postgraduate education which would see it centrally supervised by the profession itself (in the form of the GMC) in the same way that first qualifications in medicine were from 1858. The report also contained, however, an academic attack on the higher qualifications of the royal medical corporations. The report argued for the replacement of these examinations as markers of potential and actual consultant status with a system of continuous assessment of experience in hospital clinical posts. Todd was trying to locate the definition of a consultant in the hospital and university regional training structures and to distance the royal colleges from postgraduate medical education. This prompted the colleges to work together to agree common examinations as part of an alternative postgraduate structure.

The final chapter maps how the colleges' successful modification of the Todd template has formed the basic framework for the development of postgraduate education, training and examination in Britain in the last thirty years. It ends with a sketch of recent developments in medical education in Britain and how these have affected the role of the Glasgow Royal College.

On an archival note, at the time of writing the personal papers generously gifted by Dr Gavin Shaw to the College in 1997 were still in the process of being catalogued by the College Archivist, Mrs Carol Parry. They are thus referred to throughout simply as 'Shaw Papers, RCPSG 44'.

This project was made possible by the generous financial support of the Wellcome Trust and the Royal College of Physicians and Surgeons of Glasgow. We wish to record our gratitude to both of these bodies.

We would like to thank the following for their support, encouragement and guidance over the last three and a half years. At the Wellcome Unit for the History of Medicine, University of Glasgow, Dr James Bradley, Ms Jenny Cronin, Dr Marguerite Dupree, Mr Matthew Egan, Dr Campbell Lloyd, Ms Catriona MacDonald, Dr Fiona Macdonald, Ms Krista Maglen, Ms Sibylle Naglis, Ms Debbie Nicolson, Dr Malcolm Nicolson, Ms Anne-Marie O'Donnell and Ms Edna Robertson. Also in the Unit, Ms Rae Meldrum and Mrs Anne Mulholland for consummate administrative and secretarial advice and support. We also thank elsewhere in the University, Professor Anne Crowther, Mr Alistair Tough (GGHB Archivist), Professor Rick Trainor, and the staff of the Glasgow University Archives and Business Records Centre; in Edinburgh, Dr Mike Barfoot and Dr Steve Sturdy; in Aberdeen, Dr David Smith; in Manchester, Professor John Pickstone; in Stirling, Dr Jacqueline Jenkinson. At the RCPSG we would like to express a particular debt to Mr James Beaton and to Mrs Carol Parry for their constant help and support, and for both reading through the manuscript. Thanks are also due to Mrs Anna Forrest, and to the President's Secretary, Mrs Eileen Templeton. The current President of the College, Mr Colin MacKay, has been most supportive. We thank Dr Gavin Shaw for all his efforts, and also Professor Norman MacKay for his assistance. We would also like to express our gratitude to all those who agreed to be interviewed by Dr Hull: Dr Robert Fife, Dr Robert Hume, Professor Arthur Kennedy, Mr James McArthur, Professor Edward McGirr, Mr Colin MacKay, Professor Norman MacKay, Dr Gavin Shaw, Sir Thomas Thomson, Sir Andrew Watt Kay and the late Mr Stuart Young. Dr Stefan Slater was kind enough to read through the manuscript at a late stage and under great time pressure. We are grateful to the Scottish Media Group for their permission to reproduce photographs from the series 'Postgraduate Medicine', and to the Glasgow University Archives and Business Records Centre for permission to quote from their material. Finally, Andrew Hull would like to thank his wife Elisabeth for her love, so clearly demonstrated on many occasions during the writing of, to borrow from James Joyce, 'this interminable manuscript'. He would also like to dedicate this book to his father, mother and brother, with the hope that by the time she reads this his mother is better.

Andrew Hull   Johanna Geyer-Kordesch
                                                            June 1999

The College Buildings

Campbell F. Lloyd

Since the foundation of the institution in 1599, the College or Faculty Hall has occupied three known sites in the city. From 1697 the first recorded Faculty Hall building was in the Trongate, an immediate neighbour on the west side of the present Tron Church steeple.[1] Prior to this the Faculty is thought to have met in the homes of various members.[2] As part of its development in the later Georgian era, the Faculty Hall from 1791 until the 1862 was situated in a purpose built building in the then recently developed St Enoch Square. As this site became less suitable and cramped, the Faculty took advantage of its rising value as a commercial property and moved to the present site of 242 St Vincent Street.[3] As the city of Glasgow underwent economic and social changes, the College buildings reflected the developments and regenerations of the city.

The place of Glasgow, as a major port, shipbuilding, and engineering centre, in the nineteenth and early twentieth centuries, faced worldwide challenges from competition in other countries with newer techniques. Exhaustion of raw materials, new working methods and alterations in trade patterns accumulated to force change on the entire economic organisation of the west of Scotland, and on Glasgow in particular. With the decline of heavy industry on the Clyde and the agonising involved in reorientation of the economy, new industries have emerged. Part of this change has increased the value of the higher education sector as part of the economy. This brings the Royal College's activities once again to the fore as a traditionally important city institution with a view to future change, where such changes are reflected in their buildings.

While in its current home the College buildings have undergone various changes and, with an eye to the future, the Royal College has assiduously

[1] Originally the church of St Mary and St Anne.
[2] RCPSG, 1/15/17/1 A. M. Roger, 'Brief Details of the College History and College Property', p. 4.
[3] Named after the naval victory over the French off Cape St Vincent in 1807.

acquired neighbouring houses, (as recently as 1980), in part of its expansion and development programme. The present-day Hall surrounded by offices, library, examination and postgraduate facilities differs markedly from the early Faculty Halls.

When Peter Lowe was granted the Faculty Charter in 1599, Glasgow was a small cathedral city with a population of less than 10,000. Most of the streets led off the High Street and the Saltmarket and in its ribbon layout the city resembled other medieval Scottish settlements.[4] Although comparatively small, the city was home to the university, founded in 1451, and was a centre for trade in the west of Scotland, having being granted burgh status by the twelfth century. As Hollands' Britain noted in 1610, 'This Glascow is the most famous towne for merchandise in this tract. For pleasant site and apple trees, and other like fruit trees much commended'.[5]

Glasgow's historic importance was of long standing. It had been a religious site since at least the sixth century, and developed into a bishopric. Alongside St Andrew's it became the site of one of the two archbishoprics in Scotland by the Scottish Reformation in the 1560s. The city was a main fording point on the River Clyde, conveniently placed as a cross-roads along the Forth-Clyde line. The river at Glasgow was bridged in stone by Bishop Rae in 1345, although a wooden bridge had long existed prior to that. While an important site of communication, and home to archbishops; other religious administration also sustained the early city. Many of the manses and prebendary dwellings of surrounding parishes and religious houses were built near the cathedral. Even after the Reformation, in the religious disputes of the 1630s; the church general assembly met there in late 1638.

Alongside the church activities, the peripatetic royal court also had occasion to use Glasgow and charters and acts were issued from the city from the early middle ages. While the city did not have a residence directly associated with the crown, various monarchs had frequented the city before and after the Union of Crowns in 1603.[6] Various nobles, such as the Dukes of Montrose had houses or 'lands' there in the Drygate.

During the winter following the defeat of the Scottish army at the battle of Dunbar in the early 1650s, Oliver Cromwell, in his attempts to conquer Scotland, used the bishop's former country residence of Silvercraigs on the Saltmarket as his base.[7] While the city was a centre of importance for

---

[4] P. G. B. McNeill and H. L. MacQueen (eds), *Atlas of Scottish History to 1707* (Edinburgh, 1996), p. 461.

[5] J. M'Ure alias Campbell, *The History of Glasgow* (Glasgow, 1836), p. 122.

[6] McNeill and MacQueen, *Atlas of Scottish History*, pp. 132–33.

[7] *Glasgow Delineated* (Glasgow 1836), p. 17.

communications and growing trade and excise,[8] it suffered from set backs such as plague in 1647 and fire in the 1650s and 1677.[9] There were also regular floods of the River Clyde,[10] which continued well into the nineteenth century.[11]

It was as a result of fire damage that the rebuilding of Glasgow was undertaken along regulated lines with stone fronted facades and limitations on the use of timber. When Daniel Defoe saw Glasgow in the early eighteenth century, it was this planning that so struck him. Allowing for his propagandist overtones, the fine streets and stone built houses that he saw led him to describe Glasgow as one of the finest cities in the kingdom. During this period the Faculty acquired their Hall in the Trongate or St Thenews (St Enoch) Gate, which led to the West Port of the city and Shawfield Mansion, where Charles Edward Stuart held court in the city in the 1745–46 Jacobite campaign.

The Trongate Building

This first Hall, acquired in 1697, lay in the Trongate at the heart of a city centre growing with overseas trade. However Glasgow at this point still only had a population of about 13,000 and the acquisition of the Hall fell in the period of the severe economic distress of the 1690s. The Highland disturbances after the revolution destabilised the black cattle trade with England. War with France in the 1690s closed off traditional markets for the Scots made worse by growing protectionism across Europe.[12] Internally Scotland saw decline in textile manufactures, disruption of credit and by 1695 poor weather and harvests meant that famine recurrently struck parts of Scotland. Additionally the Darian Scheme caused a severe debt crisis amongst some of the commercial and landed classes,[13] although movement returned in most economic indicators by 1700.[14]

The Faculty's requirements for a meeting space in this period were relatively modest. Linked to shop premises, these provided a supplementary

[8] McNeill and MacQueen, *Atlas of Scottish History*, pp. 273–83.

[9] M'Ure, *History of Glasgow*, p. 122.

[10] *Clyde Commercial Advertiser*, 12 March 1782, noted the great height of the flood.

[11] M'Ure, *History of Glasgow*, p. 126.

[12] I. D. Whyte, *Scotland before the Industrial Revolution* (London, 1995), pp. 291–93.

[13] The Darien Scheme was an attempt to set up a Scottish trading colony in Central America. The investment for Scotland was huge, draining a third to half of the ready specie of the country. It failed, and the settlers faced a dreadful fate in the face of disease, climate, harrying attacks by Spanish and native Indian peoples, and lack of knowledge of the area and the problems any colony would encounter.

[14] T. M. Devine, *The Transformation of Rural Scotland* (Edinburgh, 1994), p. 3.

XXIV THE SHAPING OF THE MEDICAL PROFESSION

income source. Active and attuned to this by 1708, the surgeons petitioned the burgh to have a doorway to their tenement reopened and recognised as theirs; as it would be 'useful to them for enlarging of their laigh houses and chops (shops) adjoining thereto'.[15] Eventually decided in 1712, the entrance, shared with the Tron church, was opened up with the burgh keeping it as their own. Tron tacksmen and fleshers who had used the area being prohibited from either storing goods or having stands in the opening.[16] Again a petition appeared in 1737 to have the town 'give them their entry below their hall' as they sought to develop a ruinous building to the west of the site.[17] Parts of the tenement and related property would continue to yield income to the 1860s.[18]

Glasgow's population remained small but even in 1700 Scotland's total population was still only around 1,250,000, probably less than Switzerland or the city state of Venice.[19] Economic growth continued and the Faculty Hall sat near the heart of commercial Glasgow. Built diagonally across the road from the Tolbooth, where the town council met, courts could convene, and the Exchange; where trade business was transacted.[20] In an early plan of renewal and rebuilding the Faculty sought in 1757 to build a new Hall beside the Town Council Hall, which would have had the same frontage, 'mostly for ornament to the city'. Having no money they wanted the town to assist and the matter quiety disappeared.[21] Further along the north of the road was the access to the herb market in the Candleriggs with its bowling green and sugar works. While M'Ure's description of the Trongate in 1736 does not mention the Faculty Hall directly, surgeons, such as John Bogle, are noted as having their tenements in the Trongate.[22] Other businesses close by sold medicinal herbs,[23] and Faculty members no doubt were involved in the gathering of statistics for the Bills of Mortality published in Glasgow papers.[24] Newspapers contained business adverts for

[15] *Extracts from the Records of the Burgh of Glasgow, AD 1691–1717* (Glasgow 1908), p. 433.
[16] Ibid., p. 480.
[17] *Extracts from the Records of the Burgh of Glasgow, AD 1718–38* (Glasgow 1909), p. 481.
[18] RCPSG, 1/5/1, Weir MSS (1869), p. 218.
[19] B. Lenman, *An Economic History of Modern Scotland, 1600–1976* (Liverpool, 1977), p. 45.
[20] Map of Glasgow from W. Fleming merchant in Glasgow 1750–51, presented to the publisher by his grandson J. G. Fleming Esq., MD, President of the Faculty of Physicians and Surgeons Glasgow. Published by John Tweed (Glasgow, 1871).
[21] *Extracts from the Records of the Burgh of Glasgow, AD 1739–59* (Glasgow, 1911), p. 509.
[22] M'Ure, *History of Glasgow*, p. 133.
[23] *Glasgow Advertiser*, 27th January 1783, p. 2. Mrs Adams in the Blackfriar's Wynd.
[24] *Glasgow Chronicle*, 4 January 1776, p. 4, General Bills of Mortality for Glasgow, Gorbals and Anderston.

slaves,[25] cargoes and ships, such as the 'brigg' *Kattie*, sailing to trade in Tobago and the *Betsey* bound for Grenada.[26]

In this centre of trade the Faculty Hall was situated on the floor above the shops at street level. Although there is no complete visual representation of this building, it was connected to some other buildings to the back of the property and the Faculty also held property in King Street, directly round the corner.[27] The Faculty continued to own shops, flats and attics in the Trongate and cellars in King Street, valued at £2855 in 1860, while the St Enoch Hall was valued at £5000.[28] The Trongate Hall building survived as late as 1858, by then owned by the town council. By 1869 the Faculty was still in receipt of £552 from their property holdings in the old city area. They also borrowed from their Widows' Fund using it for financing until the move to St Vincent Street in the 1860s.

The old medieval city witnessed increased overcrowding as waves of immigrants from the surrounding counties, the Highlands and later Ireland congregated into the district. Subdivisions of tenements and the wynd lands were rife, with little or no investment in improvements. Even the fine houses of St Andrews Square and Charlotte Street declined in their allure as homes to the commercial classes, as later occured in St Enoch Square. Most became highly insanitary slum properties and a focus for public concern.

By the 1900s much of the remnants of the old city, around the area of the High Street, Trongate and Saltmarket, had been demolished and rebuilt by the Glasgow Improvement Trust.[29] The building which replaced the old Hall retained the two story format with crow stepped gabled windows and the city coat of arms. This format for a two story rebuilding was outlined by the city authorities as early as 1782.[30] Further pressures arose from railway developments and the site of the St Enoch Square Hall was also eventually swept away to form part of the St Enoch Railway Station.

St Enoch Square Building

The building of the new Hall at St Enoch Square reflected the development of Glasgow commercially and physically to the west. In a legal dispute

---

[25] *Ancient Glasgow* (Glasgow, 1854), p. 71, advert from 1745–47, 'To be sold, a Black Negro Boy, about five foot eight inches high, and seventeen years old. Enquire to the publisher of this paper'.

[26] *Glasgow Chronicle*, September 1776, p. 4.

[27] RCPSG, 1/5/1, Weir MSS (1869), p. 218.

[28] RCPSG, 1/13/7/9, J. Hutchison *RPFS/RCPS: Property and Environment*, pp. 40–41.

[29] W. H. Fraser and I. Maver, *Glasgow*, ii, *1830 to 1912* (Manchester 1996), chapter 11.

[30] *Extracts from the Records of the Burgh of Glasgow, AD 1781–95* (Glasgow 1913), p. 70.

between the Faculty and the magistrates and town council of Glasgow in the 1790s the Faculty had their legal right to inspect apothecaries activities confirmed.[31] However, although the Faculty was involved in a legal dispute with the city, ties remained close on other levels. In 1785 the Faculty had invested their capital stock of £1200 with the city for a 5 per cent interest return,[32] and it was the city magistrates who initially approached the Faculty over the sale of the Trongate site.[33]

As a result of the legal decision upholding their rights, Faculty members were associated with setting up an Apothecaries Hall noted in 1822 as situated in Virginia Street.[34] Working westwards by 1836 an Apothecaries Hall was listed as in Queen Street,[35] while another was in Glassford Street. These sites were only a short walk from St Enoch Square. Legal disputes involving the Faculty, with the University and individuals, cost money and valuable resources for sometimes questionable results. Cost was one of the reasons for their putting off 'the odium' of taking unlicensed practitioners to the courts.[36] As they would have been well aware, and as was seen in popular literature of the period, 'I have heard a lawsuit compared to a country dance, in which, after a great bustle and regular confusion, the parties stand still, all tired, just on the spot where they began'.[37]

The new Physicians and Surgeons Faculty Hall was set in the east side of St Enoch Square immediately off Argyll Street and cost the Faculty £1050.[38] Discussing the idea since the early 1780s, a committee worked on the proposals through to 1790, when the Trongate Hall was sold to Gilbert Shearer and Co. for around £800.[39] Described in 1824 as 'a handsome building two stories high, having a rusticated basement, supporting a range of pilasters, entablatures, and balustrade'.[40] It was erected in 1791, from designs by Mr. John Craig.[41] A more detailed contemporary description of the interior

[31] RCPSG, 1/8/5, paper on judgement, 1793.
[32] RCPSG, 1/5/1, Weir MSS (1869), p. 211.
[33] Extracts, 1781–95, p. 355.
[34] W. M. Wade, Tour of Modern and Peep into Ancient Glasgow (Glasgow, 1822), p. 196.
[35] Glasgow Delineated (Glasgow, 1836), p. 146.
[36] E. Harrison, MD, FRSAEd, An Address Delivered to the Lincolnshire Benevolent Medical Society at their Anniversary Meeting in 1809: Containing an Account of the Proceedings Lately Adopted to Improve Medical Science, and an Exposition of the Intended Act for Regulating Medical Education and Practice (London, 1810), p. 52.
[37] J. Galt, The Ayrshire Legatees (first published in 1820; republished Macmillan and Co., 1895) p. 5.
[38] Roger, 'Brief Details of the College History', p. 4.
[39] RCPSG, 1/5/1, Weir MSS (1869) pp. 216–18.
[40] Glasgow Delineated (Glasgow, 1824), pp. 24–25.
[41] Glasgow Delineated (Glasgow, 1836), p. 87.

appeared in 1822, although it wrongly listed the building as being on the west side of the square. Surgeons' Hall it noted:

> presents a neat front of two stories, and an attic; the material freestone; the lowest story rusticated; and the attic, near each end of which is a festoon sculptured in panel, surmounted by a balustrade. In the centre of the lower story, which slightly projects, is the entrance. Over this, which, as well as a window on each side, is arched; rises, from the basement to the attic, a tall window, occupying nearly the whole space between duplicated pilasters of the Doric order, and placed beneath a double arch in the attic above. This window and a shorter one on each side, light the *Hall* itself; a large, lofty, and well-finished room; containing the *Library* belonging to the *Faculty of Physicians and Surgeons*, and a fine emblematical painting, the principle figure in which is the Goddess Hygeia, with her cup and attendant serpent. Old portraits of medical worthies, hung in the lobby below, may not possess much attraction for the stranger-visitor, to whom, relative to this structure, we have only in addition to say, that within it, the *Glasgow Philosophical Society* hold or held their meetings.[42]

Craig's estimates for the costs included three marble chimney pieces and a sculptured lion and unicorn for the top of the building.[43] To the back was further accommodation for servants. The library, started in 1698, had to be housed, and from the early 1820s the nucleus of a museum was formed,[44] but the collection was eventually passed to the Infirmary on condition that Faculty members had access in the 1850s. The site looked out onto an enclosed square of grass and shrubs, dominated by the spire of the St Enoch church built in 1780.[45]

When the Faculty moved to St Enoch Square in 1791 the population of Glasgow was around 70,000 and even in 1824 there was very little built up area to the north west of the square. But Glasgow as a textile and manufacturing centre developed with increasingly rapidity and the population began to grow as the city developed trade based on these industries. The use of part of the Hall building for the vaccination programme,[46] and for other business, increased the uses on the Hall facilities and calls for a move to a new site appeared by the 1820s,[47] paradoxically at a time when their vaccination policy was in decline.

By 1821 the population was nearly 150,000. When the Faculty moved to St

[42] Wade, *Tour of Modern and Peep into Ancient Glasgow*, p. 207.

[43] RCPSG, 1/5/1, Weir MSS (1869), p. 219.

[44] Ibid., p. 353.

[45] Hutchison, *Property and Environment*, pp. 5–6.

[46] F. A. Macdonald, 'Vaccination Policy of the Faculty of Physicians and Surgeons of Glasgow, 1801 to 1863', *Medical History*, 44 (1997), pp. 291–321.

[47] RCPSG, 1/5/1, Weir MSS (1869), pp. 221, 391.

Vincent Street, in 1862, Glasgow's population alone was nearly 400,000. Crude death rates were in the region of 31.5 per thousand in the early 1830s, remaining level at nearly 30.5 per thousand to the 1860s, before they began to fall to around 25.3 per thousand by 1891 and 17.3 per thousand by 1911.[48] As the city developed in the nineteenth century, Glasgow's built up area remained highly compact, with increased overcrowding. Ticketing of tenement closes to publicly signify and limit numbers that could be legally accommodated was introduced, and night inspections had to police flatted properties to combat unscrupulous landlords. The city had a population of over 600,000 by 1901. With the surrounding burghs gradual inclusion in the city, the population was over 1,000,000 by the 1940s. In this development new mansion houses and terraces were built very near to the river and industrial areas. In the nineteenth century it has been noted that 'older institutions' such as the Faculty, 'experienced a dramatic upsurge in size and importance'.[49] As such the pressures on the St Enoch site increased.

The second Hall was eventually sold in 1862 to William Teachers & Sons and later they sold the property to the Glasgow and South Western Railway Station for the St Enoch Hotel. The Faculty considered moving at various points from 1810 onwards and in 1846 to 1847 a committee report recommended a site in Cambridge Street to the north of Sauchiehall Street. However the only offer of £3760 received in 1847 faced rejection as being low when the buildings had been valued at £4500. As Weir notes, by 1860, 'it brought a thousand pounds more'.[50] Paradoxically, Glasgow University moved to Gilmorehill by the late 1860s and its former site on the High Street also became a railway station.

St Vincent Street Building

The Faculty moved from the site at St Enoch Square in 1862 to an new home in present-day St Vincent Street at No. 242.[51] The area was then called Blythswood Hill forming part of the Blythswood estate offices of the Campbell's of Blythswood.[52] The area developed in the 1800s, with streets, such as Bath Street named after the great water cisterns there, being laid out. The ground was feued into plots and the district beyond initially set up as pleasure

[48] Fraser and Maver, *Glasgow*, ii, p. 147.

[49] Ibid., p. 232.

[50] RCPSG, 1/5/1, Weir MSS (1869), p. 392.

[51] St Vincent Street was bisected to the east by Waterloo Street.

[52] For an account of the development of the Georgian and Victorian city see the introduction by E. Williamson, A. Riches and M. Higgs, *The Buildings of Scotland: Glasgow* (London, 1990).

[53] Hutchison, *Property and Environment*, p. 12.

gardens 'on the lines of Vauxhall'. These were near to some small mansions and the botanical gardens opened in 1819.[53] Change rapidly transformed the area so by the time the Faculty moved to the site it was dominated by Alexander 'Greek' Thomson's United Presbyterian church built in 1858, the pleasure gardens having disappeared under new streets and buildings.

No. 242 was larger than its neighbours and after committee reports on the suitability of the building for the Hall and other needs, it was bought in June 1862 for £4500, from the Rev. George Ann Panton.[54] From the National Trust for Scotland's viewpoint it forms part of St Vincent Street as,

> the remains of another, different symmetrical terrace of the late 1830s. In the centre, the five-bay No. 242, with fluted couple columns and a balustrade to the central portico, was a superior house altered by Sir J. J. Burnet of the Royal Faculty (now Royal College) of Physicians and Surgeons in 1892–3. He panelled the staircase hall (which still has its screen of Tuscan columns, though the sequence of saucer domes looks typical of Burnet) and rebuilt the staircase to form a grand axial approach to the College Hall, which he added behind the house above ground-floor classrooms (now altered). A subtle change of rhythm on the top-lit mezzanine landing before the hall itself, a rather bleak double height room, its coved plaster ceiling with a large guilloche pattern and its big timber fireplace and doorcase in late seventeenth century style. Across the front of the building at first floor level, the drawing room of the 1830s house, the plasterwork on walls and ceiling apparently original. Burnet was responsible for the careful modelling on this landing, which, until altered by W. N. W. Ramsay in 1962–63, was lit by a fine dome above the octagonal gallery that served the second floor. The dome can still be seen from the second floor library, which was extended across the light-well. All the post-1914 alterations have been quite without Burnet's imagination. On the ground floor, the original dining room, with Corinthian-columned screens and a galleried library, handsomely fitted up by Burnet, was spoilt in 1956 by the mania for stripping varnish. It is now entered from No. 236, a narrower house taken over in 1901 and slightly altered by Burnet. He may have put the gallery in the first-floor drawing room, still with part of the fine 1830s ceiling. All the houses have original staircases and columned halls.[55]

By 1869 the Faculty had undertaken redecoration of the Hall, library and reading room and bought new furniture.[56] Costs remained a factor, as for example in the illuminations to celebrate the marriage of the Prince of Wales and Princess Alexandra in 1863. Paraffin lamps being cheaper and more

[54] Ibid., p. 53.
[55] Williamson, Riches and Higgs, *Buildings of Scotland*, p. 237.
[56] Hutchison, *Property and Environment*, p. 41.
[57] Ibid., p. 57.

reliable were used instead of an expensive electrical device for their illuminated coat of arms.[57]

The Faculty, while employing well-known architects like Burnet to make
alterations to the buildings, kept Fellows and Members abreast of costs and
new building practice with library purchases such as the G. Lister Sutcliffe's
two-volume *Principles and Practice of Modern House Construction* (London,
1899).[58]

As Hutchison notes, much of the present interiors of some of the houses
hide the original building layouts. More unfortunately, many of the original
plans for the buildings are missing.[59] Nevertheless Burnet's alterations sig-
nificantly improved the utility of the houses to the Faculty. As a living
building, the Royal College has continued to make alterations through to
the present day. Many are related to the increased teaching load of the College
such as for the Maurice Bloch Lecture Theatre plans of 1959,[60] a development
based on the then gift of £20,000.[61] The Fraser Library development of 1963,
while obscuring the dome for the stair well, created a fine working space for
College members and consulting readers and a lift installed in 1959 has
improved access.

During the late 1970s and early 1980s Glasgow underwent a dramatic
change in two senses. Huge levels of rebuilding work destroyed many of the
city's Georgian and Victorian buildings and terraces as the city 'fathers and
mothers' aimed to make Glasgow the world's most modern city by the year
2000. Adam-designed buildings, built as private houses in College Street and
Charlotte Street, disappeared, and within sight of the present College buildings, a motorway cut through the Victorian city. A regeneration programme
began with stone cleaning of city buildings hidden under layers of domestic
soot and industrial pollution. The College buildings inclusion in this regeneration has created a delicate balance between conservation and
practicality.

The Royal College complex in St Vincent Street has recently undergone
another major refurbishment. The Maurice Bloch Lecture Theatre has been
completely upgraded and the Crush Hall made into a more comfortable
waiting area for anxious candidates awaiting College examinations.

[58] RCPSG, Lib GC. E. 18.
[59] Hutchison, *Property and Environment*, p. 22.
[60] RCPSG, 1/6/11–13.
[61] Hutchison, *Property and Environment*, p. 130.

Future expansion of medical knowledge will undoubtedly tend to make medical education an increasingly selective process for every student. This is not a new problem, however: the point has long been passed at which one person could have more than a superficial knowledge of all areas of medicine ...

The social position of the doctor himself is also liable to change in important respects. As progress in science and technology continues, attitudes towards doctors, as towards members of other professions, are likely to move still further in the direction of regarding them as experts to be called in to prevent, investigate and remedy specific functional defects rather than as members of an elite who are accorded a special status by virtue of their general background and qualifications. The very fact that the doctor is concerned with the most personal aspects of human health, and indeed with the fundamental matters of life and death, will ensure a continuing high prestige for his profession; but the esteem in which the doctor is held by the community in general will be determined much more by his demonstrated competence than by the mystique of his calling.

Report of the Royal Commission on Medical Education, 1965–68, Cmnd 3569, (London, HMSO, 1968), p. 30, para. 30.

After the Medical Act, 1858–1886

[Dr R. McDonnell] You think you would rather remain as you are?

[Dr R. Scott Orr, President, FPSG] We would rather remain as we are.

If the question were not would you rather remain as you are; but would you rather be disestablished? – Extinguished you mean.

Yes? – We would rather remain as we are.

But I say if you were told that you were not going to be allowed to remain as you are, and the alternative put before you were, that progress, or what is called progress, now requires that a certain change should be made, and you had your choice of alternatives, either of being extinguished or of falling in with the conjoint board scheme, which would you think best?-Of course we would fall in with the conjoint scheme, and take our place upon it which we would consider ourselves entitled to.[1]

The 1850s were a time of successful consolidation for the Glasgow Faculty during which it secured its own independent existence and privileges against the earlier threat of absorption by the more prestigious and richer Edinburgh Colleges.[2] First of all, the passing of the Faculty Act in 1850 gave a new legal

---

[1] *Report of the Royal Commission Appointed to Enquire into the Medical Acts*, Parliamentary *Papers* (PP), 1882, xxix (C. –3259.-I.), p. 106.

[2] The key secondary literature for this period includes A. M. Cooke, *A History of the Royal College of Physicians of London*, iii (Oxford, 1972); Zachary Cope, *The Royal College of Surgeons of England* (London, 1959); James Coutts, *A History of the University of Glasgow, 1451–1909* (Glasgow, 1909); W. S. Craig, *History of the Royal College of Physicians of Edinburgh* (Oxford, 1976); Tom Gibson, *The Royal College of Physicians and Surgeons of Glasgow: A Short History Based on the Portraits and Other Memorabilia* (Edinburgh, 1983); S. W. F. Holloway, 'Medical Education in England, 1830–1858: A Sociological Analysis', *History*, 49 (1964), pp. 299–324; Irvine Loudon, *Medical Care and the General Practitioner, 1750–1850* (Oxford, 1986); idem, 'Medical Practitioners 1750–1850 and the Period of Medical Reform in Britain', in Andrew Wear (ed.), *Medicine in Society* (Cambridge, 1992), pp. 219–47; idem, 'Medical Education and Medical Reform', in V. Nutton and R. Porter (ed.), *The History of Medical Education in Britain* (Amsterdam, 1995); Charles Newman, *The Evolution of Medical Education in the Nineteenth Century* (London, 1957); N. and J. Parry, *The Rise of the Medical Profession* (London, 1976);

imprimatur to the Faculty's status under the 1599 charter.[3] In a move which aligned the Faculty with the emerging lay and professional groundswell towards national legislation on medical qualifications, the Faculty publicly forfeited its rights to exclusive local control of licensing. In return it gained recognition of its qualifications as equal to, and 'conferring the same Status and Privileges' as, those of any other Scottish medical corporation.[4] This was not just local equality in Glasgow and the four counties,[5] but now extended 'throughout Her Majesty's Dominions'.[6] The Faculty Act thus established the Faculty as a player in the ongoing medical reform debate, and as a progressive one wedded to the modern concept of standardised national qualifications, as championed by John Simon the Medical Officer to the Privy Council in London. Secondly, the Faculty had gained, after tenacious negotiating, its place at the top table of official British licensing bodies by being named as such in Schedule A of the 1858 Medical Act.[7]

This double success understandably led to a mood of self-satisfied optimism within the Faculty. In 1860 it was decided that the old city had become too dirty and crowded to be a suitable place to house the newly recognised national medical corporation, and the St Enoch's Square Hall, which had been the Faculty's home since 1791, was put up for sale. It was bought in 1862 by William Teacher and Sons for £5500, and was finally demolished in 1875, when the site was redeveloped by the Glasgow and South Western Railway for the construction of a station and hotel. The city was growing westward, and the Faculty followed this trend and relocated to 242 St Vincent Street, which was purchased from the Reverend George Ann Panton, a Minister of the Free Church of Scotland, on 7 April 1862 for £4500.[8] Panton had previously

F. N. L. Poynter (ed.), *The Evolution of Medical Education in Britain* (London, 1966); T. N. Stokes, 'A Coleridgean against the Medical Profession: John Simon and the Parliamentary Campaign for the Reform of the Medical Profession, 1854–8', *Medical History*, 33 (1989) pp. 343–59; see also Stokes's thesis, 'The Campaign for the Reform of the Medical Profession in Mid-Victorian England' (unpublished M. Phil. thesis, University of Cambridge, 1987); Ivan Waddington, 'General Practitioners and Consultants in Early Nineteenth-Century England: The Sociology of an Intra-Professional Conflict', pp. 164–88, in J. Woodward and D. Richards (eds), *Health Care and Popular Medicine in Nineteenth-Century England* (London, 1977); also his *The Medical Profession in the Industrial Revolution* (Dublin, 1984).

[3] *An Act for Better Regulating the Privileges of the Faculty of Physicians and Surgeons of Glasgow and Amending their Charter of Incorporation*, for full text see Alexander Duncan, *Memorials of the Faculty of Physicians and Surgeons of Glasgow, 1599–1850* (Glasgow, 1896), appendix 6, pp. 229–31. Hereafter referred to as the Faculty Act.

[4] Ibid.

[5] Lanark, Renfrew, Dumbarton and Ayr.

[6] Faculty Act, ibid., p. 229.

[7] *Bills Public: The 1858 Medical Act* (21 and 22 Victoria, cap. 90).

[8] 'History of the College Hall', R. B. Wright, RCPSG, 44 (Shaw Papers).

run a school for young ladies on the site. This move was soon to be echoed by the University, which was also drawing up plans in the 1860s to leave its old town site in the midst of the eighteenth-century commercial district and decamp to Gilmorehill, a relocation that was finally completed in 1870. The original College (University) site had become surrounded by low-quality housing and by chemical and other works. A flavour of the insalubriousness of the old town area by the 1860s was given in a contemporary report from the Scottish University Commissioners which described the College as:

> surrounded with an atmosphere impregnated with the effluvia arising from the filth occasioned by such a population, in a town of which the sewerage is far from being in a satisfactory condition, and with the fumes and vapours of the chemical and other manufactories [and] it is hardly possible to conceive a combination of circumstances less favourable to the bodily and mental well-being of the youth attending a University, or less suitable for conducting the business of a public seminary of instruction.[9]

St Enoch Square, once a residential area for doctors, had now become a thriving commercial district, meaning both that the Faculty property was now of optimum value, and that there was much disturbance from crowds and traffic.[10] The new St Vincent Street premises were in an up-and-coming area of the city. The Faculty's new neighbours on the opposite side in 1860 were eight merchants, a minister, a soldier, one superintendent of Ordnance Survey, a house factor and a milliner. The magnificent St Vincent Street Church, designed by Alexander 'Greek' Thomson and built in cream-coloured stone, had only recently been completed in 1859.[11]

In November 1862, at the Annual Meeting (AGM) of Faculty Dr John Gibson Fleming [12] suggested that the Faculty should also now adopt 'more

[9] *A Report of the Buildings of Glasgow University*, appended to *The Report of the Scottish Universities Commission* (1863), p. 210. This quotation is reproduced from J. D. Mackie, *The University of Glasgow, 1451–1951* (Glasgow, 1954), p. 280.

[10] John Hutchison, *RFPS/RCPS: Property and Environment*, c. 1980, RCPSG, 1/13/7/9.

[11] Ibid. Happily the church is to be restored in time for Glasgow's year as City of Architecture and Design in 1999.

[12] John Gibson Fleming (1809–79), MD Glasgow 1830, Fellow of Faculty in 1833. He became surgeon to the GRI in 1846, served on the GMC representing the Faculty from 1863–78, and was twice President of the Faculty (1865–68; 1870–72). With Watson he was instrumental in securing the Faculty Act of 1850. He was Manager of the GRI (where he improved the contracts of the staff), the Royal Asylum for Lunatics, the Eye Infirmary and the Maternity Hospital, and prime instigator of the GRI Medical School (later St Mungo's College) founded in 1876, after the University had taken all its clinical teaching to the Western Infirmary. He served for thirty-five years as chief medical adviser to the Scottish Amicable Life Assurance Society, whose head office was in Glasgow. His main published work was *The Medical Statistics of Life Assurance*. Details from Gibson, *RCPSG*, pp. 149–51.

classical and heraldic' armorial bearings, in keeping with its newly confirmed 'dignity and position'.[13] After consideration, the Council reporting on 6 April 1863 (four months after the move to the new buildings) declared that the existing emblem (thought to date from 1657 and depicting Asclepius's rod with entwined serpent, with a poppy for physicians and a lancet for surgeons and the motto, *Conjurat Amice* – in amity we live):[14]

> certainly expresses very plainly our connection with the healing art but does not indicate that we are a corporation by virtue of a Royal charter of great antiquity ratified by Act of Parliament and consequently an important national as well as local institution. In the opinion of Council therefore the Faculty as an ancient National Scientific Corporation and as one of the licensing bodies of the United Kingdom ought in justice to itself to have such an emblazonment as will denote in a more heraldic and classic style its objects, Royal foundation, nationality and locality.[15]

The new coat of arms consisted of a central quartered shield which contained in the first and fourth quarters the original Asclepian emblem; in the second quarter, bearing testament to the original charter from James VI in 1599, the royal arms of Scotland; and in the third quarter the arms of the city of Glasgow. The shield was supported on the right by a figure of Minerva, goddess of Science and Art, and on the left by a figure of Hygeia, the goddess of Health. The traditional motto was retained, and a new one, *Non Vivere sed Valere Vita* (not simply to live but to enjoy life), added.[16]

However, this mood of self-satisfaction was not to last. While the 1858 Act had assured the Faculty's status as a national licensing body, the legislation had left much confusion as to which qualifications examined the candidate in *both* medicine *and* surgery. The profession favoured a self-regulating system, and the proposed extensive powers of the General Medical Council to set curricula and examinations had been drastically diluted. While this left discretionary powers and much independence with the licensing bodies, the absence of a standardised national qualifying system in medicine (the single portal of entry to the profession) meant that all licensing bodies had to pay attention to what the market for general practitioners (for instance the Poor Law service or the Army) required and it increasingly demanded

[13] *Faculty Minutes*, 3 November 1862, RCPSG, 1/1/1/9. I am grateful to James Beaton, the College Librarian, for drawing my attention to this and the following quotation. See also Hutchison, *Property and Environment*.

[14] The old motto comes from Horace, *Ars poetica*, c. 4.11. Details here from, 'Brief Details on College History and College Property', n.d., author unknown, RCPSG, 44, pp. 6–7.

[15] 'Report of Council on New Armorial Insignia of the Faculty', *Faculty Minutes*, 6 April 1863, RCPSG, 1/1/1/9.

[16] The new motto was a quotation from Martial's *Epigrams*, book 6.

comprehensive qualifications that demonstrated the individual's ability to practise medicine and surgery. The London Colleges struggled to put together a conjoint board offering a comprehensive qualification throughout the 1860s, but the task was even more vital for a small corporation like the Glasgow Faculty, whose qualifications could easily be overlooked by candidates wishing to take no gamble with their medical futures who thus opted for only the most prestigious British qualifications. An additional factor was the national campaign led by John Simon to achieve his goal of legislation for the single portal. Thus it made economic sense in this period for the Faculty to attempt to formally ally itself with the Edinburgh Colleges to grant comprehensive qualifications.

Debate and agitation about legislative reform of the education and legal definition of the medical profession was far from over with the passing of the Medical Act in 1858. Rather, the Act, and the experience of its working over the next ten years, brought the unresolved questions into sharper focus. The Act's principal Architect, the Medical Officer to the Privy Council, John Simon, had set out to impose a single portal of entry to the profession through common national examinations. However, mindful of the interests of the medical authorities, Simon's chosen way of achieving this was subtle. The authorities would continue to award their various qualifications, but these would be supervised by a powerful General Medical Council (representing the corporations, universities and government) with authority to compel adherence to a standard curriculum and examination. This Council would be able to block sub-standard qualifications from being registrable, compel combination of medical authorities for examination, or set its own examinations.[17] However, Spencer Walpole (the new Conservative Home Secretary in the Derby/Disraeli administration which had assumed office on the fall of Palmerston's ministry) influenced by the corporations and by his own gradualist approach to reform, had scotched the clause giving the GMC such wide powers. It was replaced by a clause giving the Council only the right to inspect examinations and obtain information on curricula. The GMC could appeal to the Privy Council to stop qualifications being registrable if they were, in its opinion, sub-standard, or to impose a general curriculum or examination change which it considered essential.[18] In practice, however,

[17] See Simon, 'Memorandum Prepared in 1858 by the Medical Officer of the then General Board of Health (Mr Simon) in Explanation of the Medical Practitioners Bill of that Year, as Drawn for the Board under Mr Cowper's Presidency', reprinted in *Parliamentary Papers*, 1879 (320), Special Report of the Select Committee on the Medical Act (1858) Amendment No. 3 Bill [Lords], appendix 1, pp. 305–10. See also ibid., p. 310, for Simon's original wording of the clause giving the GMC wide powers.

[18] See ibid., pp. 310–11, for the amended clauses in the Act reducing these powers.

up to 1886, this was never done lest such action cause schism and deadlock in a Council which, after all, was made up of representatives of those medical authorities whose examinations would be the subject of criticism. Thus as Royston Lambert has observed: 'Simon's concept of a Medical Council specifically empowered to reshape medical education on one-portal principles had shrunk into the reality of a Council without any effective powers over education at all'.[19]

During the early to mid 1860s, Simon was busy with other projects (for instance the introduction of compulsory qualification in vaccination for every medical practitioner under the Vaccination Act of 1867), and the medical authorities were at least attempting to raise standards and unite for examinations. But, by late 1867, frustrated by the slow pace of improvement, the failure of the GMC to use its right of appeal to the Privy Council to force the pace of change, and the continued existence of so-called half-qualifications, which were granted after examination in medicine *or* surgery but not on the full range of medical knowledge needed for general practice, Simon once again began to act. It was his initiative which produced the extended continuation of the national debate about the medical profession and its education which rumbled on from 1870, through twenty-five Bills, two Select Committees and a Royal Commission, and culminated in the Medical Act Amendment Act of 1886.

The whole second period of intense debate about medical reform in Britain is perhaps even less written about than the first, which produced the Medical Act of 1858. There was, of course, merely one greatly extended debate which periodically became more, nationally and politically, heated. The proximate causes of the increased temperature between 1870 and 1886 (and the particular hotspots of 1870, 1879, 1883 and 1886) were largely the manoeuvrings of Simon. This time he had attempted to orchestrate even more carefully the process of reform. Simon helped to create a renewed public debate, got the GMC to ask him for a single portal to be legislated for, consulted and then drafted the Bill himself, and, of course, prevailed on the government to introduce it. Although it was finally shelved forever by the 1886 Act, nevertheless Simon's single portal must have seemed to have, if not an irresistible, than certainly a dramatically increased momentum throughout much of this period.

This chapter will examine the Faculty's role in these national developments, assess their implications, and analyse the Faculty's continual manoeuvrings to ensure that, once again, it maintained its existence, rights

---

[19] Royston Lambert, *Sir John Simon, 1816–1904, and English Social Administration* (London, 1963), p. 464.

and privileges in changed circumstances. It is important to understand that the loci of meaningful action had now changed. After 1858, because of the way in which Simon forced the question onto the government's immediate policy agenda, all meaningful action occurred at the national level: between Simon and the GMC (Simon goading the Council into action, and it sounding out opinion and cooperating in framing government Bills); and to a lesser extent between the GMC and its regional branch councils (English, Scottish and Irish). The Faculty had representatives on both the national General Medical Council and its Scottish Branch Council,[20] but, in both forums, it was only one small voice among many. Furthermore, again showing the intense national level of debate in medical circles, much of the decisive action came in Parliament: Lords and Commons. Here the medical interests lined up to enlist support in amending or defeating key Bills (and this was doubly complicated by the fact that medical interests, in the form of the GMC, had often cooperated in the drafting of these Bills with Simon). Finally there was the general political climate: this was not 1858 and the generation of politicians who wanted reform but not revolution in the familiar and traditional British institutions had largely gone, to be replaced by men who were at least familiar with a more radical reforming context. It was the difference between the parliamentary generation which passed the Great Reform Act in 1832 and that which had very recently passed the Second Reform Act in 1867.[21]

All this serves to underline the fact that the Faculty's scope for action in this period of the medical reform debate was much more limited. It was very much on the periphery of a focused national process of reform which was being negotiated in London, mostly at a high level. Its ability to influence the shape of reform was, correspondingly, even less than it had been in the lead up to 1858. Like other old medical authorities, the Faculty was thus concerned to preempt (and thus hopefully delimit) national moves by reforming itself voluntarily (usually in combination with the Edinburgh Colleges). If the authorities were seen to be capable of voluntarily putting

[20] The Faculty's representatives in this period were from 1858–1860, James Watson; from 1860–63, George Watt; from 1863–78, John Gibson Fleming; and from 1878, Robert Scott Orr. All were eminent Faculty members and Glasgow medical men. With the exception of Watt, all were already, or became, President of the Faculty during their membership of the GMC. These same men also represented the Faculty on the Scottish Branch Council, which was usually composed of all the Scottish GMC representatives plus Scottish Crown Nominees such as Andrew Fergus. See *Faculty Minutes* (passim) and *Parliamentary Papers*, 1879 (320) appendix 11, p. 425.

[21] On this see, for instance, Asa Briggs, *Victorian People* (Harmondsworth, 1971; originally 1954).

their house in order, there was perhaps less chance of the state depriving them of professional control for the public good.

However, the confusion caused by the 1858 Act as to which qualifications examined in all aspects of medical knowledge (medical and surgical) added a further impetus to voluntary combinations of medical authorities. If they offered double qualifications or even triple qualifications (that is, Simonian combined boards of corporations and universities), they would be better able to attract candidates wishing to enter the public services, and thus better ensure their own survival. In short, the marketplace for medical knowledge was itself pushing in the direction of more comprehensive centrally administered examinations in each of the three divisions of the Kingdom.

The Faculty was also, as ever, in competition with the universities (also engaged in a parallel process of internal reform via the Commissioners of the Scottish universities) to provide the cheapest examination that was acceptable to the greatest number of employers of medical expertise. After an initial renewal of old hostilities between the Scottish corporations and the universities, the interests of the two merged together. Radical reform was staved off once and for all partly out of deference for the case made by the Scottish universities against inclusion in a London-orientated national scheme, and the strength of their support in Parliament.

The Faculty would have preferred that nothing changed. But realising that some form of single portal, which would undermine their independent authority, prestige and possibly even financial survival, seemed clearly visible on the legislative horizon, the Faculty, like the other British medical authorities, was determined to obtain the most benefit from the new conditions of operation. This situation is indicative of the position that the Glasgow Faculty (and later College) often found itself in the period covered by this book: adjusting to changed national circumstances as the major professional players and the state negotiated the reform of the medical profession. The very role of the medical corporations was constantly changing as self-regulation and professional status interacted with the increasing trend towards state control of medical education and, ultimately, practice.

The Double Qualification

The combination of one or more medical authorities for the purposes of examination had been explicitly sanctioned by clause nineteen of the Medical Act. The Faculty had long considered some kind of union with the Edinburgh Colleges to strengthen its position in the national context by becoming part of a Scottish College of Surgeons. For example a scheme had been extensively discussed in 1856, at the height of the debate that led to the Medical Act. A distinguished Faculty delegation, consisting of the ex-Presidents James Watson[22] and Alexander Dunlop Anderson,[23] attended the first meeting of the Scottish Branch of the General Conference of the Corporations of the United Kingdom, in Edinburgh on 21 November 1856. On the table were two plans for the administration of the proposed new conjoint Scottish board for licensing general practitioners. The first proposed an amalgamation of the Faculty with the Edinburgh surgeons to form a new body – the Royal College of Surgeons of Scotland – with two divisions: one in Glasgow and one in Edinburgh. The Faculty title was to be given up and the new Glasgow body would be called the Royal College of Surgeons of Scotland in Glasgow. Importantly, a new charter bearing this name was to ensure its relative autonomy: it was to be able to hold property in this title 'and enjoy the immunities and discharge the duties at present enjoyed or discharged by the Faculty'.[24] Each city division was to appoint equal numbers of examiners to the board with diets being held in each city. Fellows of Faculty who wished to join the Edinburgh Physicians as a result of this arrangement were to be admitted *ad eundem* to the RCPEd, and the Physicians were to supply assessors to the conjoint exam board to make up one third of its complement.

The second plan involved no change to the current names or organisation of the three Scottish corporations, but merely suggested that they combine to constitute a 'General Licensing Board for Scotland', each body contributing an equal number of examiners and diets being held in both cities.[25]

The branch conference, having elected a President and Secretary (Drs Wood and Maclagan respectively) and secretaries for each city (Maclagan for

[22] James Watson, later styled 'the Father of the Faculty' for his part in the Faculty Act and in the negotiations securing the Faculty's place in the Medical Act of 1858, was born in Glasgow in 1787. He became a member of Faculty in 1810 and graduated MD at the University of Glasgow in 1828. After serving as a Surgeon at the GRI he was appointed physician there in 1842, and was also physician to the Clyde Street Fever Hospital. He was Faculty President at three critical periods, 1838–41, 1849–52 and 1857–60, and was also the Faculty's first representative on the GMC. See Duncan, *Memorials*.

[23] Alexander Dunlop Anderson was born in 1794 and graduated MD from Edinburgh in 1819. He was nephew of Dr John Anderson, founder of Anderson's College; see John Butt, *John Anderson's Legacy: The University of Strathclyde and its Antecedents, 1796–1996* (East Linton, 1996). He was admitted MRCS Eng. in 1816 and became FRCS in 1844. He entered the Army Medical Service, and was assistant surgeon to the 49th Regiment. In 1820 he settled in practice in Glasgow, he entered the Faculty as a Fellow in 1821, and in 1823 was appointed surgeon to the GRI, and, in 1838, physician there. He was President of the Faculty from 1852–55. See Duncan, *Memorials*.

[24] 'Report of the Delegates from 18 June till 1 December', *Faculty Minutes*, 1 December 1856, RCPSG, 1/1/1/8.

[25] Ibid.

Edinburgh and Watson for Glasgow), discussed the two plans. The Edinburgh surgeons objected to both, and Wood suggested a third way: no title changes, no conjoint exam board *but* uniform exams by each city's surgeons with assessors attending from the Edinburgh Physicians. The Faculty objected to this plan, and the question was referred to full meetings of each body to come up with a decision.[26] The Faculty's final opinion was to reject Wood's plan but to support the operation of either of the first two.[27]

At a further meeting of the Scottish branch in early February 1857 it was agreed that for the present the Faculty and the RCSEd should continue to operate as separate licensing authorities for general practitioners, with the involvement of the RCPEd, which would provide two assessors to each board and attest to examination in medicine on the reverse of the diploma certificate. The curriculum for the diploma was to be the same for both boards, and practical midwifery, natural history and comparative anatomy were to be added. There would also be a common preliminary examination before entering professional education. This would include English composition, Latin, mathematics, and natural philosophy, and candidates would also be encouraged to study German, French and Greek.[28]

These negotiations finally faltered on the degree of autonomy to be accorded the Glasgow arm of any combined body. Clearly, the Faculty was not willing to subsume its identity in a new body necessarily dominated by the more powerful Edinburgh Surgeons. However, in 1859, when the Faculty was digesting the implications of the Medical Act, a very real threat to the sufficiency of its qualifications spurred the Glasgow body into renewed negotiations with Edinburgh.

During early 1859 it became clear that two of the most important markets for the employment of Faculty Licentiates (the Poor Law Board and the Army Medical Department) had altered their standards of entry and now demanded two separate qualifications: one in surgery and one in medicine. While the Medical Act had created a register of qualified practitioners, the continuing variety of registrable qualifications, and the legality of registering with qualifications in either medicine and surgery combined *or* singly,[29] left a legacy of uncertainty in the minds of employers of medical expertise as to the extent of knowledge of the whole practice of medicine possessed by individual candidates.

---

[26] Ibid.

[27] *Faculty Minutes*, 1 December 1856, resolution proposed by George Watt and seconded by Dr Tannahill.

[28] *Faculty Minutes*, 2 February 1857, report of delegates (Watson and Anderson) from Scottish Branch meeting, Edinburgh, 21 January 1857.

[29] See Medical Act (1858), clause 19.

The Faculty had received a plea for help in early 1859 from seven Scottish qualified physicians and surgeons acting as Board Medical Officers in the North of England (one of whom – Andrew Morison – was a Licentiate) complaining that a recently promulgated order of the Poor Law Board of London stated that no Medical Officer could hold any permanent parochial appointment under the Board unless he possessed two qualifications (one in medicine and one in surgery).[30] In spite of the notional equality of all qualifications recognised as registrable under the Medical Act (and, before this the highly-prized guarantee of equal treatment for the Faculty licence under the 1850 Faculty Act) employers of medical expertise were clearly making independent judgements of the quality of different qualifications. Furthermore, there was an English bias against Scottish corporate and university qualifications which were perceived as narrowly-based and easier, and were not considered sufficient (either singly or in tandem). Such bias is only understandable as a reaction against the large number of Scottish-trained doctors who practised in England (usually after taking an additional English qualification). Scottish medical education was more academic, as the medical schools had traditionally been (largely) based on universities. This contrasted with London, where the medical schools were based on the great teaching hospitals. As William Tennant Gairdner later noted, this meant that before the Medical Act, Scottish doctors had tended to be more well-versed in basic sciences though less experienced at prescribing and other practical applications; whereas London educated doctors had tended to have wide clinical experience, but less knowledge of chemistry and anatomy.[31] The Medical Act had brought England into line with Scotland by abolishing the Apothecaries Company's monopoly on the training of the general practitioner, thereby forcing the English Colleges to cater for him as well as elite physicians and surgeons. Apprenticeship was thus replaced by systematic and clinical teaching in the hospitals. In Scotland even the apprenticeship system: 'was always controlled by a regular academic discipline, emanating largely from

[30] See *Faculty Minutes*, 14 February 1859, RCPSG,1/1/1/8. The matter was also referred to the Scottish Branch Council of the GMC, see *GMC Minutes* (SBC Section), 1, pp. 272–73.

[31] W. T. Gairdner (1824–1907), one of the most important figures in the Victorian Glasgow Medical School. After appointments in Edinburgh as first pathologist (1848) and then physician (1853) to the Royal Infirmary, he succeeded Professor Macfarlane as Professor of Medicine at Glasgow University. In 1863 he also became the city's first Medical Officer of Health (he was succeeded in 1872 by Dr James Burn Russell). Gairdner was an avid scientific naturalist in the Huxley mould and his medical work reflected an enduring attachment to the use of the latest aids in clinical diagnosis. See G. A. Gibson, *Life of Sir William Tennant Gairdner* (Glasgow, 1912), which includes reprints of many of his medical and general articles and addresses.

the Universities, but ... reinforced ... by the coordinate influence, and often rivalry, of some of the corporations in building up what is now called the medical curriculum'.[32]

Systematic and clinical teaching had replaced apprenticeship much earlier in Scotland, and the University and medical corporation curricula covered medicine and surgery to produce qualifications for general practice. In Glasgow, the University Medical School curriculum was first broadened by the progressive influence of William Cullen and Joseph Black as successive Professors of Medicine (1751–1766), and then by the opening of the Glasgow Royal Infirmary (GRI) to patients in December 1794 to provide a reservoir of clinical material for students.[33] Nevertheless, clinical teaching remained in an unsatisfactory condition into at least the 1830s. There was only medical clinical teaching by the Infirmary physicians who were all FPSG Fellows or Licentiates and were appointed by the GRI Board and the FPSG. The introduction of clinical teaching in surgery, vocally desired by students was complicated by disputes between the Faculty and the University over who should appoint clinical teachers, which in turn was part of the wider 'long lawsuit' between the two over who had the right to license surgeons. Clinical instruction in both departments of medicine by the infirmary staff was initiated in 1829 but was still poor and resulted in the University losing students to Anderson's College of Medicine, which added eight new medical lecturers between 1819–40, and enrolled more students than the University between 1840–60.[34]

Such a clear picture of the true strengths and weaknesses of Glasgow or Scottish medical education had, however, clearly escaped the London-based Poor Law Board. All the Scottish and Irish universities and corporations were affected by the Poor Law Board's Consolidated General Order fixing the new qualifications for Medical Poor Law Officers, and most of them sent memorials directly to the board in London demanding that their particular qualification be accepted for entry as covering the whole range of medical knowledge.[35] In response the board involved the GMC, and the

---

[32] W. T. Gairdner, *The Physician as Naturalist* (Glasgow, 1889), appendix to chapter 1, 'Extract from Introductory Address Delivered in the University of Glasgow, to the Students of Medicine of the Session, 1882–3', p. 58.

[33] See Derek Dow and Michael Moss, 'The Medical Curriculum at Glasgow in the Early Nineteenth Century', *History of Universities*, 7 (1988), pp. 227–57.

[34] See ibid., and Jacqueline Jenkinson, Michael Moss and Iain Russell, *The Royal: The History of the Glasgow Royal Infirmary, 1974–1994* (Glasgow, 1994), pp. 74–78.

[35] The bodies involved were the universities of Edinburgh, Glasgow, Aberdeen (one each from Marischal and King's; there being no united university till later in the year) and London, and the two Edinburgh Colleges, the Faculty and the Irish Surgeons. See *GMC Minutes*, 1 (1858–60), 10 August 1859, pp. 65–66.

communication is revealing. The board stated that it had originally debarred Scottish and Irish degrees in medicine as it believed that they were not legally recognised as entitlements to practise in England. Such was the confusion accompanying the Medical Act that it was only now that the board admitted that all British degrees were now legally recognised throughout all British territories. The board also argued that it had always insisted on candidates possessing a qualification which fully covered the practice of medicine, pharmacy and surgery. How far each particular qualification did this was, the board argued, still open to doubt, and it requested a ruling from the GMC:

> the important question still remains, as to what is the exact extent and nature of the qualification which is obtained by the Degrees or Licences conferred by the several Bodies who have made their applications to the Board. The Board do not possess any authoritative information on the subject which they can act upon; and they therefore request to be informed by the General Medical Council, how far the Degrees, Diplomas or Licences, of the several Bodies ... confer respectively, the right of practising Medicine or Surgery, or Medicine and Surgery, and are evidence that the persons to whom they are granted have attained a competent knowledge of either, or both, of those branches of the profession.[36]

On petitioning the Poor Law Board itself, the Faculty was told that it was delaying recognition of the Faculty's (and other Scottish) qualification(s) until it had checked their sufficiency with the GMC.[37]

The GMC worked quickly on this crucial problem, announcing their findings at a special meeting on 10 August. The GMC report, however, did not clarify matters. It referred the board back to the confusing clause 31 of the Medical Act, stated that all registrable qualifications had suitably broad curricula, and merely listed the type of qualification (degree or licence, in medicine, or in surgery), which the original memorialists had been granted. The Faculty was thus listed as offering a licence in surgery, although this was, of course, a legal qualification for *general* practice.

The GMC had not given an adjudication on a tricky point in the Act, because it did not believe this was within its power. But its attempt at application of the Act, already demonstrated its shortcomings. The market had already moved beyond the Act, in wanting to know curriculum details and insisting on a broadly examined candidate with a clear double qualification, because the Act was not perceived as adequate public protection against medical ignorance.

---

[36] W. G. Lumley (Assistant Secretary, Poor Law Board) to the Secretary of the GMC, 5 August 1859, reproduced in *GMC Minutes*, 1 (1858–60), pp. 50–52.

[37] See *Faculty Minutes*, 7 March 1859 and 4 July respectively.

This new tendency was underlined by the Faculty's discovery of a similar policy in operation in the Army. In a letter to the FPSG President, James Watson, in July 1859, the Director General of the Army Medical Department had stated that:

> every candidate, before being allowed to present himself for the competitive examination for the Army Medical Service must produce a degree in Medicine, or a Licence to practice it from some body corporate in Great Britain or Ireland legally entitled to grant such, *in addition* to a Diploma in Surgery or Licence to practise it from some body in Great Britain legally entitled to grant such Diploma or Licence. I beg to add that the object in requiring a qualification from two distinct bodies was to raise the standard of education in candidates for the Army Service, and I am of the opinion that this can best be obtained by requiring the candidate to produce the Double qualification.[38]

The Medical Act, and perhaps particularly the emasculated and cautious type of GMC it created, was then not sufficient even to stop discrimination between legally registrable qualifications. The Faculty, as ever, moved quickly to find a practical solution, despite the fact that the Faculty licence was a broad qualification in this period which included examination in the whole range of medical practice (See Table 1 for the full curriculum for the single licence at this time). But swift action was needed to stop the licence – the basis of the Faculty's financial survival and continuing prestige – becoming obsolete.

Having discerned similar problems about the sufficiency of their qualifications with potential employers, the two Edinburgh Colleges had begun negotiations between themselves only in June on the granting of a Double Qualification (DQ) embracing medicine and surgery. In Edinburgh it was the Surgeons who had initiated contacts. A joint qualification with the Physicians, which would be publicly known as a 'double' qualification, would very visibly extend the range of their nominally surgical licence to include the full sweep of medicine. This kind of arrangement was exactly what the Faculty needed to shore up the public image of its own (again nominally surgical) licence, and it approached the RCPEd at the end of June asking to enter into a similar arrangement with the Physicians. A first meeting was set up early in July, and its announcement in the *Faculty Minutes* underlined the proximate cause of such action. Junction with the Physicians was in view, 'as to the examination and granting of diplomas to Licentiates in order to meet the requirements of the Army and Navy Boards and the Poor Law

---

[38] Ibid., 4 July 1859, letter from J. Alexander, Director General, Army Medical Department to the President of the FPSG, James Watson, 21 June 1859. My emphasis.

Table 1. *Curriculum for the FPSG Single Licence and the Double Qualification of the FPSG/RCPEd*[39]

1 Anatomy	Two Courses	Six months each
2 Practical Anatomy	One Course	Twelve months
3 Chemistry	One Course	Six months
4 Practical or Analytical Chemistry	One Course	Three months
5 Materia Medica	One Course	Three months
6 Institutes of Medicine (Physiology)	One Course	Fifty lectures
7 Practice of Medicine	One Course	Six months
8 Clinical Medicine	One Course	Six months
9 Principles and Practice of Surgery	One Course	Six months
10 Clinical Surgery	One Course	Six months
11 Midwifery and the Diseases of Women and Children	One Course	Three months
12 Medical Jurisprudence	One Course	Three months
13 Practical Pharmacy	One Course	Three months
14 Hospital Practice		Twenty-four months
15 Dispensary Practice		Six months
16 Morbid Anatomy		Three months
17 Practical Midwifery		Six cases

Double Qualification

Candidates must take one three month course in Botany and a further course of six months in either Practice of Medicine or Clinical Medicine and Surgery or Clinical Surgery.

Fees

The fees in 1861 were £10 for the single licence and £16 for the Double Qualification.

[39] See curriculum as detailed at the end of George Buchanan's, 'Remarks on the New Medical Act, as Affecting the Curriculum of the Student: An Address to Students Delivered at the Opening of the Medical Session, 1861–62' (Glasgow, 1861), in RCPSG, Glasgow Collection, *Glasgow Scientific Pamphlets*, 14, 21–22.

Commissioners'.[40] Whatever the exact new requirements of the various public employers, they seemed to the Faculty to enshrine the principle that
'candidates for their service are to hold Diplomas from two separate Licensing
Bodies'.[41]

Further meetings on 23 July in Glasgow and on 25 July in Edinburgh of
the Council plus other eminent Fellows with the RCPEd Council (which
included notables such as the President and GMC representative Dr Andrew
Wood, and William Tennant Gairdner) followed. The proposals proved
remarkably easy to draw up and agreement too was swift. But then there
was a clear sense of urgency and the already existing deal between the
Edinburgh bodies to work from. The alternate locations of the meetings
underlined that this was a limited agreement for practical reasons and that
the Faculty preserved complete autonomy. This point was further underlined
by the RCPEd's assurance that Physician Fellows of the Faculty resident in
Glasgow would be admitted to the Edinburgh College as Fellows, 'in the
most liberal manner'.[42] Watson had also obtained assurance that such Fellows
of both bodies living in Glasgow would be preferred when the RCPEd was
selecting its examiners for the double in Glasgow.[43] While they would be
bona fide Fellows of the RCPEd, they would also have a prior (and continuing) existence as Fellows of Faculty; their examination fees would easily
cover the (no doubt moderate) cost. All the Edinburgh Physicians wanted
was a share of the fees.

While this arrangement was far from underhand, after all Glasgow Physicians were perfectly entitled to join the Edinburgh Physicians, and were
qualified to do so, and the Faculty curriculum did cover a good deal of
medicine as it stood, its convenient arrangement does perhaps provide insight
into the mode of voluntary agreements of this kind. It was this unregulated
nature that made Simon uncertain. Something of a Doubting Thomas when
it came to the corporations (and particularly the Scottish ones), he wished
to see double qualifications firmly regulated to ensure their comprehensive-
ness, with a central curriculum and examination administered by joint
boards.

The Scottish arrangements were approved by the GMC (it should be noted,
after somewhat frenzied activity at the Faculty to rush the documents through

[40] Faculty Minutes, 2 July 1859, RCPSG, 1/1/1/8.
[41] Ibid., in report of minutes of conferences with RCPEd.
[42] Watson had written to Wood on 27 July asking if, as part of the Double Qualification
agreement, such persons would be admitted 'on easy terms and ad eundem' to the Fellowship
of the RCPEd. Faculty Minutes, 5 September 1859, RCPSG, 1/1/1/9.
[43] Ibid.

in time for approval by the next GMC meeting) on 8 August.[44] The regulations allowed for the curriculum to be taught not only by universities but also by extra-mural colleges whose teachers were members of a Schedule A body (in Glasgow any Fellow whose lectures were Faculty approved). Again underlining the point that the Faculty perceived its existing (single) licence to cover medicine as well as surgery, the mandatory curriculum for the DQ added only a three-month course in botany and a second six month course of practice of medicine or clinical medicine.[45] Successful candidates would receive two separate diplomas (LRCPEd and LFPSG), signed by office-bearers from each body, and thus registrable under the Medical Act as two separate qualifications.[46]

In early August the Faculty learnt that the GMC had replied to the Poor Law Commissioners that the Faculty was legally only entitled independently to confer a surgical qualification. However, by means of the DQ it was now able to also confer a complete qualification in medicine and surgery, theoretically retaining the same examiners as for the surgical licence, and with a very similar curriculum. The Faculty could and did continue to award its own surgical licence, or single qualification. This was still useful for those candidates not thinking of entering the public services, but destined for 'ordinary private practice' as a GP.[47] The substantial difference between single and double qualifications for the candidate was cost: £10 and £16 respectively.[48] The new DQ was publicised by newspaper advertisements under the names of both the Edinburgh College and the Glasgow Faculty.

The Faculty and the Scottish Universities

While the corporations had been responding to pressure from the public employers of medical expertise for a double qualification via voluntary arrangements, the universities had been undergoing a parallel process of official state reform to provide for more comprehensive medical qualifications.

The Scottish universities were similarly troubled by the new requirements

[44] *Faculty Minutes*, 1 August 1859, RCPSG, 1/1/1/8, reports and minutes of meetings of 23, 25, 26 and 29 July.

[45] See Table 1 for the curriculum as detailed at the end of George Buchanan's *Remarks on the New Medical Act, as Affecting the Curriculum of the Student: An Address to Students Delivered at the Opening of the Medical Session, 1861–2* (Glasgow, 1861), in RCPSG, Glasgow Collection, *Glasgow Scientific Pamphlets*, 14, pp. 21–22. Buchanan was Lecturer on Anatomy at Anderson's Medical School (as well as a Fellow of Faculty and surgeon at the GRI), and this address was to the Andersonian medical students.

[46] Regulations for Joint Licence, *Faculty Minutes*, 5 September 1859.

[47] Buchanan, *Remarks on the New Medical Act*, p. 13.

[48] Ibid., p. 21.

of the public services for a double qualification in 1859. The GMC announcement of 10 August had left public doubt whether a university degree
in medicine, though legally acceptable for general practice, was actually also
a mark of surgical knowledge. Glasgow University had been listed by the
GMC in August as offering separate qualifications in Medicine (MD) and
Surgery (CM). Thus, while both qualifications legally qualified holders to
practise, in the climate of concern about the breadth of medical knowledge
signified by individual qualifications, both of these could appear to employers
as partial qualifications. The concern was not at the level of the curriculum
required for individual qualifications, but at the title conferred. The nomen-
clature of Scottish medical degrees thus needed to be simplified, and (together
with the curricula), standardised. This was the task which the Scottish University Commissioners, given executive power to reform by the 1859 Universities
(Scotland) Act, set about between 1859–62. The commissioners have been
portrayed as imposing an English model of specialisation essentially alien to
the tradition of broadly-based Scottish higher education. Oxford and Cambridge had recently instituted successful honours schools and triposes,
respectively, and this was, partly, an attempt to extend this kind of specialisation to Scotland.[49] However, this specialisation was also meant to attack
real problems with university qualifications, such as the status of medical
qualifications noted above. It may have been alien to the Scottish educational
tradition, but the Scottish universities did have to respond to the demands
of the broader, national (British), market for medical knowledge.[50]

The Commission began by drawing up a new medical examination plan
for Edinburgh University. The MB (Bachelor of Medicine), which had been
first introduced at King's College, Aberdeen in 1852, was established along
with a new degree of CM (Master of Surgery). Both degrees demanded the
same curriculum and examination, the candidate could chose to take the
MB without the CM, but not the other way around. The MD degree was
made a higher qualification after the MB.[51] This gave Edinburgh the same
legal entitlement to license in surgery as Glasgow. The long dispute between
the Faculty and Glasgow University from 1826–40 had ended with the House
of Lords agreeing with the Privy Council's ruling that the Faculty must license
every Glasgow surgeon. But the right of the University to offer its own
qualifications in surgery was not disputed, although the popularity of the

[49] See A. Lockhart Walker, *The Revival of the Democratic Intellect Scotland's University
Traditions and the Crisis in Modern Thought* (Edinburgh, 1994), chapter 4, especially pp. 66–68.
[50] On these reforms, see also R. D. Anderson, 'Scottish University Professors, 1880–1939:
Profile of an Elite', *Scottish Economic and Social History*, 7 (1987), pp. 27–54.
[51] Details here are from Coutts, *History of the University of Glasgow*, pp. 570–77.

CM plummeted after 1840. The Medical Act's equalising of all corporate and university qualifications might have seemed to have made battles about the nature of the relative remit to examine and grant qualifications obsolete, but, in July 1860, the two Edinburgh Colleges, the Faculty and the English Surgeons now legally challenged the commissioners' ordinances with the Queen in Council. Meeting with no success there, the dispute was taken up with the Privy Council in January 1861.[52] The fear was that if universities were able to offer double qualifications, these would be preferred by students to their less prestigious corporate equivalents.[53]

This was essentially a renewal of the 1815 Court of Session case that the Faculty had brought against Glasgow University arguing that only licentiates could practise surgery (and therefore be general practitioners) within Glasgow. It was only the confusion of the Act and the uncertainty of the public employers that, by creating insecurity among the medical authorities and a perception that they must once again fight for their survival in the medical marketplace, allowed this renewed challenge to emerge.

The whole tendency of recent medical reform, unclear though it was, was against the exclusive rights of the corporations, and their challenge was finally dismissed, and the Edinburgh ordinances approved by the Queen in Council on 4 February 1861. The way being thus cleared, similar regulations were applied to Glasgow and (the now unified) Aberdeen. The new regulations meant that candidates for the MB at Glasgow had to attend at least one year at that university and a further year at another recognised university. However, in a concession to the corporations, the Commissioners made 'large and liberal' provisions for the recognition of the courses of extra-mural teachers as counting towards the degree.[54] In Glasgow, this gave a stimulus to Anderson's Medical School with which the Faculty had very strong links. It supplied the majority of the staff, who were usually also physicians or surgeons at the GRI (through which students thus had access to clinical teaching on payment of a class fee), and many of the Andersonian students took the Faculty's examinations.

[52] Abstracts of these cases are given in 'Scottish Universities Commission', *Edinburgh Medical Journal*, 6 (1861), pp. 765–73.

[53] For the procession of memorials and petitions in this dispute see *The Medical Corporations versus the Scottish Universities, 1859–61*, a bound collection of the relevant ephemera preserved in RCPSG, Glasgow Collection.

[54] Coutts, *History of the University of Glasgow*, p. 572, quoting the Commissioners.

National Developments

The Scottish corporations and universities had both shown how the unwieldy and confusing Medical Act could be worked to provide comprehensive, publicly identifiable broad qualifications. Simon was, however, not confident that such piecemeal action was the most efficient way to organise medical education. He still desired a clear national system of qualification to sweep away all the uncertainties. Apart from his desire to impose a centralised order, he did not trust the medical authorities to reform themselves, believing that there were too many vested interests, and that medical education was too important to be left to the medical elite.

In 1867 Simon renewed his reform agitation.[55] Late in the year, as part of concerted manoeuvrings to create a reformist climate, he became President of the Medical Teachers Association, a reform pressure group. On 20 January 1868 he delivered a strong speech critical of the corruption and abuses of the medical corporations and of half-qualifications. Simon as MO to the Privy Council was not in a constitutional position to initiate reform, the demand had to come from the profession itself. But there was nothing to stop him from helping things along. He thus also got himself elected on a reform ticket as a member of the Council of the RCSEd and began to stir up the Fellows.[56] Caught up in the pressure for reform among its Fellows thus created, the College proposed (via the GMC) to the Privy Council a government Bill making minor amendments to the Medical Act. This was just the opportunity Simon had been working for. He wrote to the GMC on 14 May 1869 stating that the Lord President was not willing to introduce a government Bill that did not cover 'all the ground where amendment of the Medical Act is wanted',[57] and requesting the cooperation of the Council in drawing up such a wide-ranging Bill. Simon identified two particularly pressing problems. First, the Act did not provide a 'satisfactory and uniform standard of admissibility to the Medical Register'. Secondly, the Act had not sufficiently empowered the GMC 'to issue regulations in this respect'. He

[55] The argument in this section is indebted to Lambert, *Sir John Simon*, especially chapters 19 and 24, and to the 'Statement in Regard to Recent Attempts at Medical Reform' in appendix to *Parliamentary Papers* (1882), pp. xxxi–xivi.

[56] Before his career in public office began in 1848 with his election as first Medical Officer of Health for the City of London, Simon was a surgeon at King's College and St Thomas's Hospitals in London. At the latter he was apprenticed to his mentor J. H. Green and rose to Lecturer in Pathology at St Thomas's Medical School in 1847. He received an Honorary Fellowship of the RCSEd in 1844. See Lambert, *Sir John Simon*.

[57] Simon to President of the GMC, 14 May 1869, reproduced in *Parliamentary Papers*, 1879 (320), 'Special Report from the Select Committee on the Medical Act', appendix 1, pp. 313–14.

noted with alarm that thousands of registered practitioners had only half qualifications, and while admitting that most medical authorities extended their curriculum beyond the titular limits of the qualification, doubted, 'whether that mode of action, at its best, can supply more than a very imperfect substitute for complete legal qualification'.[58]

The GMC had already appointed a sub-committee to look at further reform. Its report agreed with Simon on the problem of half qualifications. It noted that the GMC had recommended to the medical authorities that all registrable qualifications should include medicine *and* surgery, and that the authorities had moved in this direction (with new or combined examinations) but concluded that, 'the only adequate remedy for this acknowledged defect would be for the Council to accept, under an amended Medical Act, such powers as would enable them in the future to refuse registration to any person, whatever his legal qualification may be, who has not passed sufficient examinations both in Medicine and Surgery'.[59] Simon had got half of what he wanted from the GMC, the rest came in the report of its education sub-committee which Simon received at the beginning of August. This report endorsed his idea of joint boards for each division of the kingdom, stating that:

> One of the great evils at the present moment, is the inequality of the examinations for the licence. This inequality of the test of efficiency is the more unfortunate, as every licence confers an equality in the right to practice everywhere. The easy examination of one licensing body tends to depress the standard of the examination in all the rest. Visitations of examinations doubtless partly remedy this state of things, but, to completely remove it, a bolder course is necessary. The time has now arrived when, leaving to universities and corporations full liberty to deal as they please with their honorary distinctions and degrees, the Medical Council should endeavour to effect such combinations of the licensing bodies included in Schedule A as may form a conjoint examining board for each division of the kingdom, before which every person who desired a license to practise should appear, and by which he should be examined on all subjects. Any higher degrees he may wish to take should come after, and should be optional.[60]

The chairman of this committee and author of the report was Andrew Wood, the President of the RCSEd and their representative on the GMC. His support for compulsory combination of medical authorities into joint boards now is important, as later he was to be one of its strongest and most effective opponents.

[58] All quotations here from Simon's letter to the GMC, ibid., p. 313.

[59] 'Letter from the President of the GMC, with Resolutions of Council, and Report of Committee', 17 July 1869, ibid., p. 315.

[60] Extract of report of GMC Education Committee, ibid., p. 318.

Simon followed up his favourable position with a more explicit letter to the GMC in February 1870. In it he criticised the existing examination system as run by mutually independent and competing authorities, each with their own registrable standard, and stated that if the GMC wanted a radical reform Bill it should get the issue mandated by a clear vote. On 26 February the GMC voted seventeen to one in support of joint boards in each division (the *Faculty Minutes* commented knowingly that 'in the state of public and professional feeling no other course was open to them').[61] Simon, having expertly manipulated the situation and created a feeling that the single portal was the only solution, drafted the radical Bill.

The government Bill of 1870 was introduced into the House of Lords in April by the Lord President, Lord Ripon. In its original form it was 'a ruthless embodiment of the one-portal principle',[62] laying down that a double qualification was needed for registration and that conjunction of the medical authorities in each division of the kingdom into three joint examining boards was now compulsory. Only the examinations of the joint boards were to be passports to registration. The joint board schemes were to be vetted by a GMC newly empowered to standardise the curriculum and supervise the examinations so that the licence would be granted on the same terms throughout the United Kingdom. The medical authorities were to cease to grant their individual licences, but might grant their membership to joint board licence holders without examination or fee. The Medical Register could include a column for the listing of higher titles. In an effort to recompense the medical authorities for loss of revenue, surplus fees from the joint board exams would be shared for the upkeep of libraries, museums and examination halls.

This Bill got a favourable reception from the medical authorities and was approved by the GMC. However, in its passage through Parliament it was amended in favour of the universities. The clause debarring authorities from awarding titles to anyone but joint board licensees was omitted and a new clause was inserted allowing acceptance of examinations in fundamental medical science subjects of other bodies. The Faculty, in tandem with the other two Scottish corporations, strongly objected to these amendments. The Faculty petitioned the House of Lords in favour of the original Bill, arguing that only the primacy of the joint board licence could end half qualification throughout the nation, as the pioneering double qualification had ended it in Scotland. The amendments, it was argued, 'would destroy the uniformity of the Scheme, and ... would render it completely inoperative in remedying

[61] *Faculty Minutes*, 11 July 1870, RCPSG, 1/1/1/9.
[62] Lambert, *Sir John Simon*, p. 470.

the principal defect of the Medical Act of 1858'. Acceptance of university qualifications in lieu of joint board ones would shatter the scheme, and the granting of medical degrees as honorary titles to eminent men (the basis of the university case for continued authority to grant independent degrees) would be harmful to the public good by creating two classes of practitioners (registered and unregistered). This would also make prosecution of irregular practitioners extremely difficult. The Faculty's final objection to the Bill was perhaps the most important: it pressed the case for compulsory affiliation of successful joint board candidates to one of the medical authorities. As the medical authorities were to be entirely deprived of their licensing functions, they had to find a new role. The Faculty thus argued for compulsory affiliation portraying the authorities as guardians of professional and moral standards.[63]

The Faculty and the Edinburgh Colleges also produced a joint statement against the objectionable clauses,[64] and sent a joint deputation to London (the President, John Gibson Fleming, going for the Faculty), before the Bill went into committee, to exert pressure for the original Bill. This met with no success, but did stop further unfavourable amendments by the Duke of Richmond.[65]

However, the Bill eventually had to be withdrawn on 25 July 1870. It had run into implacable opposition in the Lords over the increased powers of the Privy Council under the Bill (Lord Salisbury had argued that he was 'not prepared to abolish the nineteen licensing bodies for the purpose of installing Dr Simon in their place'),[66] and faced opposition in both Houses from supporters of the British Medical Association. The BMA backed changes to the composition of the GMC to include representatives directly elected by the profession at large, a point upon which Simon was unwilling to budge.

Attempts at Voluntary Reform

The government did not introduce another medical Bill for eight years after the failure of 1870, though Private Members put forward over twenty Bills up to 1881, none of which were successful in achieving reform on the key

---

[63] 'Petition of the Faculty of Physicians and Surgeons of Glasgow to the House of Lords in Regard to the "Medical Act (1858) Amendment Bill"', 13 June 1870. RCPL Archives, miscellaneous papers labelled 'Medical Reform Boxes'.

[64] 'Statement by the Royal College of Physicians of Edinburgh, the Royal College of Surgeons of Edinburgh and the Faculty of Physicians and Surgeons of Glasgow in reference to the Medical Act 1858 Amendment Bill', *Minutes of the RCSEd*, 16, 27 June 1870, pp. 211–19.

[65] *Faculty Minutes*, 11 July 1870, passim.

[66] *Hansard*, cci, col. 262. Cited from Lambert, *Sir John Simon*, p. 473.

issues of examination and registration, and the composition of the Medical Council.[67] However, the events of 1870 provided a stimulus to voluntary combination of the medical authorities into joint boards. The GMC was keen to achieve what it could in spite of the lack of legislation, and the authorities were keen to reform *themselves*, and thus render unfavourable legislation unnecessary.

On 25 July 1871 the GMC issued a circular to the medical authorities urging conjunction. In England negotiations for a joint scheme had been going on since 1870 and were now renewed between the RCPL, RCSEd and the universities of Oxford, Cambridge, London and Durham, and the Society of Apothecaries. This process rumbled on through false starts and abandoned agreements until at last a much reduced scheme including only the two English Colleges was finally established in October 1884.[68]

In Scotland similar efforts were made. On 31 October 1871, 'considering the demand which exists for uniformity in examination',[69] the Faculty met in conference with the two Edinburgh Colleges and agreed general proposals as the basis of arrangements for a conjoined board for Scotland. The scheme proposed was a compromise which, while instituting a joint board, preserved the maximum degree of autonomy for the existing medical authorities. The examination of the proposed board would be the only portal of admission to the Medical Register as Licentiate in Medicine and Surgery, but, in an effort to attract the universities into the scheme, it was only to examine in practical (clinical) subjects. Echoing the pro-university amendments to the 1870 Bill, the first examination in the fundamental medical science subjects (anatomy, physiology and chemistry) was to be left with the existing medical authorities. However, they agreed not to award their degrees or licences to anyone who had not passed the board's final exam. Further resolutions were passed at a meeting of the three corporations on 21 November: a twelve man committee of management with equal university and corporate representation was to be established to run the new examination. This would fix the number of examiners, the times and places of examinations and the division of subjects.[70] The corporations first met with the universities to discuss the new board on 27 December, and after further corporate meetings in the New Year, the details of the scheme were fleshed out. It was proposed

---

[67] For a list of these unsuccessful Bills see *Parliamentary Papers* (1882), appendix, schedule, pp. xlvii–xlix.

[68] For details see, Cooke, *History RCPL*, pp. 850–60, passim.

[69] 'Resolutions Agreed to by Delegates from the RCPEd, RCSEd and FPSG at Conference. Held at Edinburgh on 31 October', *Faculty Minutes*, 6 November 1871, RCPSG, 1/1/1/10.

[70] *Faculty Minutes*, 4 December 1871.

that the existing examinations of the medical authorities should continue but that the examinations in clinical medicine and surgery should be taken over by the board. The board would only examine those who had passed the final exam of a Schedule A body, but no qualifications would be granted by the Scottish medical authorities until the board's exam had been passed. The examiners were to be elected by the corporations and universities.[71]

In further meetings of the three corporations, and with the Scottish Branch Council (SBC) of the GMC (at which representatives of the universities were present), four different schemes were presented and discussed.[72] The main point at issue was the extent of the examinations of the new joint board: whether it should examine in theory as well as clinical practice. After a further intervention from the GMC urging a speedy conclusion,[73] arrangements were finalised between all the Scottish medical authorities at a conference with the SBC on 24 May 1872. The parties agreeing to participate were the three corporations and the universities of Edinburgh, Glasgow, Aberdeen and St Andrews. The scheme now laid down that the joint board would examine in clinical medicine, clinical surgery and midwifery, and that only candidates who had completed the standard GMC curriculum of medical study would be admitted to examination. The participants agreed only to grant their qualifications after the candidates had passed the examination of the board. The SBC was to act as the committee of management and examinations would be held in Edinburgh, Glasgow and Aberdeen, examiners being chosen equally by the cooperating bodies (St Andrew's only being allowed one).[74]

While the corporations remained keen, the governing bodies of the major universities decided to reject the scheme.[75] Representatives of Glasgow and Edinburgh feared that standards would be lowered by the replacement of university qualifications with those of the joint board.[76] The unspoken fear was that if the joint board licence was the portal of entry to the profession, there would no longer be any need for students to take the more prestigious university courses. As long as the GMC curriculum was covered, students could take cheaper, extra-mural classes, thus depriving the universities of status and revenue. This problem was particularly acute in the final form of

71 *Faculty Minutes*, 5 February 1872.
72 See *Faculty Minutes*, 6 May, 3 June and 1 July 1872; see also evidence of Andrew Wood before the Select Committee in 1879, *Parliamentary Papers* (1879) (320), questions 4049–54.
73 See *Faculty Minutes*, 1 April 1872.
74 Details of the scheme are reproduced in *GMC Minutes*, 10 (1872–73), 9 July 1872, pp. 9–11; and in *Faculty Minutes*, 3 June 1872.
75 See *GMC Minutes*, ibid., pp. 11–13.
76 See evidence to Select Committee of William Turner, *Parliamentary Papers* (1879) (320), question 3317; and of John Struthers, *Parliamentary Papers* (1880) (121), question 32.

the conjoint scheme where candidates did not have to pass, but merely to study, the fundamental medical science subjects (where the universities had a high public reputation for excellence) before being examined by the board. The universities' position was ably summed up by John Struthers, Professor of Anatomy at Aberdeen before the Select Committee in 1880. The Scottish universities perceived their contribution to the standard and character of Scottish medical education to be too important to be put at risk by participation in a joint board:

> as a result of our Scottish university system ... the medical profession occupies a higher status than it does in England, that they are more educated men, and with less of the apothecary about them. We have never had an Apothecaries Company to lead us to charge by medicines and the like. That is the result of our Scottish Universities; I know of no country where the general practitioner is better educated than in Scotland.[77]

William Turner, Professor of Anatomy and Dean of the Faculty of Medicine at Edinburgh University, added that there was an essential difference between English and Scottish universities that made a comprehensive joint board impossible in Scotland. He argued that the Scottish universities were the only bodies in Britain which offered a complete, high standard, medical education leading to a degree. Medical teaching in England still (in spite of reforms) revolved largely around the London teaching hospitals. Thus the Scottish universities had more to lose from a joint scheme, and so did the nation in terms of properly educated doctors:

> look at the profound difference as regards medical education between the Scottish Universities ... and the English Universities. The Scottish Universities are great graduating and teaching bodies, not only for Scotland, but also largely for the British Empire. I do not think, at this moment, there is a single university in England which can give a complete medical education, so that they are in an entirely different position. The English Universities may yield some of their existing privileges for the sake of carrying out a conjoint board scheme, and actually give up but little; whereas if we were to yield we should give up rights and privileges, and interests and duties which we feel that we cannot part with ... We are educating bodies, whereas the corporate bodies in England are not ... they are examining bodies; but we are in the double position of being both educating and examining; in fact, we fulfil in Scotland completely the ancient function of a university; a function which is fulfilled all over continental Europe by the universities as teaching and examining bodies. We carry out the theoretical idea of a university, and we reduce it to practice.[78]

[77] Evidence of Professor John Struthers before the Select Committee, *Parliamentary Papers* (1880) (121), question 83.
[78] Ibid., evidence of William Turner, questions 3244 and 3246.

This double function (teaching and examining) of the Scottish universities was one reason why, as Sir James Paget argued,[79] there was a much greater rivalry between the Scottish universities and corporations, than in England. The small size of the population, and the fact that two of the major universities were in direct competition with the corporations located in the same cities, also increased the intensity of the competition for students. This rivalry would continue until 1948 when the extra-mural schools were absorbed by the universities and the university degree finally became the *de facto* single portal of first entry to the profession. The universities then completely took over undergraduate teaching and examining for home students (although the Triple Qualification – TQ – the establishment of which in 1884 is discussed later, and the old single licence remained as possible first qualifications for overseas students). As we will later see, this development left the Faculty in crisis, looking for a new role.

There was also another continuing aspect to the rivalry between the universities and the medical corporations. The tension between the two was not only about attracting the greater number of students, but was also bound up with a particular academic conception of medical knowledge, which changed over time, but which was initially essentially based on (inherited, elite) medical theory and a general liberal education, as opposed to experience of clinical practice. In Glasgow, this latter conception of medical knowledge as experiential was traditionally and enduringly associated with the Faculty, with its easier access to clinical teaching at the GRI.

Influenced by developments in the pre-clinical sciences, in the 1870s and 1880s a new generation, of what Malcolm Nicolson has termed 'clinician-scientists',[80] emerged – mostly in university posts – and began to develop a methodology of clinical research in the hospitals, which incorporated scientific advances (in diagnosis and then therapeutics), but was founded on the critical combination of science with clinical acumen, and derided the narrow focus of laboratory science. In Glasgow we might point to university professors like Gairdner as characteristic of this type. For instance, recalling the 1870s, Gairdner commented in 1901 that, 'a professor of medicine ... must be a hospital physician; and that his hospital work and instruction form the one and only form of laboratory instruction which can be attached to his

[79] Evidence of Sir James Paget to the Select Committee, *Parliamentary Papers* (1879), 320, question 2453.

[80] Malcolm Nicolson and David Smith, 'Science and Clinical Scepticism: The Case of Ralph Stockman and the Glasgow Medical Faculty', paper presented to Science and Technology Dynamics Internal Progress Conference, Amsterdam, 18–19 September 1997. I am grateful to the authors for sight of a copy of this paper.

chair so as to keep his systematic instruction by lectures vivid and fresh and well up-to-date'.[81] Thus Gairdner resembled not only Nicolson's characterisation of the 'clinician-scientist', but also Christopher Lawrence's elite London Physicians who incorporated elements of the new methodology of laboratory science into their clinical practice, while insisting on the epistemological primacy of clinical acumen.[82]

Academic medicine, in its most progressive aspects, thus changed to mean medicine strongly informed by, though (at least among the clinical, as opposed to the pre-clinical academics) not necessarily subservient to, the techniques and methods of laboratory science. This aspect of the rivalry – competing changing conceptions of medical knowledge – also continued throughout the twentieth century. When the universities achieved a monopoly in undergraduate education after 1948, the increasingly important sphere of the postgraduate education of the hospital doctor was still largely based on clinical experience, and was the preserve of the medical corporations. The later parts of this book trace how the postgraduate education and training of the hospital doctor – like his clinical practice itself – became increasingly scientised and specialised to more closely resemble the new academic medicine – in a word, academicised.

In the 1870s and 1880s the pre-clinical sciences were featuring more and more heavily in the undergraduate medical curricula of both universities and corporations (via the GMC).[83] We will see some of the Faculty's responses to this pressure later. But, just as the Faculty needed to incorporate more science, so the University of Glasgow was keen to provide a more developed clinical element to the medical education it offered. After the University moved westwards to Gilmorehill in 1870, the Western Infirmary (GWI), which opened to patients in 1874, was the purpose built clinical reservoir for University medical clinical teaching. Between 1870–74 students had been transported to the GRI in fleets of omnibuses, but with the opening of the new teaching hospital all clinical teaching was transferred to the GWI. The Professor of Medicine (William Tennant Gairdner) and the Professor of Surgery (George H. B. MacLeod) transferred their appointments to the GWI and now taught clinical classes there. Two new clinical chairs were also

---

[81] See his 'Memories of College Life: Especially in the Sixties and Early Seventies', p. 60, in *The Book of the Jubilee: In Commemoration of the Ninth Jubilee of the University of Glasgow, 1451–1901* (Glasgow, 1901), pp. 42–68.

[82] Christopher Lawrence, 'Incommunicable Knowledge: Science, Technology and the Clinical Art in Britain, 1850–1914', *Journal of Contemporary History*, 20 (1985), pp. 503–20.

[83] See Stella Butler, 'Science and Medicine in the Nineteenth Century: Changing Conceptions of Clinical Practice', conference paper, Science in Modern Medicine, Manchester, April 1985.

created attached to the new infirmary. Thomas McCall Anderson got the new medical chair, and George Buchanan the surgical. The proximity of the new hospital to the new University site was obviously the prime factor in this switch but the earlier history of difficult relations between the University and a GRI dominated by the FPSG (through staff loyalties and the membership of its board of management) may also have been a consideration. The access of University staff to clinical teaching had long been a bone of contention, as we noted earlier. The conduct of clinical teaching at the Royal was also less than satisfactory to progressive university professors like William Tennant Gairdner, who were committed to the academic ideal of the essential unity of teaching and research and limited private practice.[84] Gairdner described clinical teaching at the Royal just before the move thus:

> There was ... a peculiarity in the arrangements which was far from satisfactory, and which shews how little organisation had at this time entered into the idea of clinical teaching. No physician and no surgeon of the Royal Infirmary had, at the time I am speaking of, what could be called a clinical *class*. The pupils entered their names in a book, and signed from time to time their attendance, not on any particular teacher, but simply as pupils of the Infirmary at large. Two of the four physicians, and a like number of the surgeons, were told off in rotation to give the instruction by two 'clinical lectures' in the week; and, for the rest, all was confusion or happy accident, every pupil going to the wards just as much or as little as he pleased, and without reference to any particular course of tuition at the bedside – the very seat of true clinical teaching. It was impossible, on this happy-go-lucky plan, to feel that, either as regards the students or their teachers, the great resources of the Royal Infirmary were being fully appreciated or wisely employed.[85]

Gairdner noted that regularised arrangements for conducting clinical teaching had to be rather painfully negotiated at the new infirmary. Clearly not all clinical teachers shared his modern views of the critical importance of clinical instruction.[86]

This development left the GRI in a very difficult position: there were now very few medical students remaining. As John Gibson Fleming, who as well

---

[84] As well as the evidence of the previous quotation from Gairdner, he also believed that a professor should not conduct private practice because it interfered with his teaching duties. See *Report of the Royal Commissioners Appointed to Inquire into the Universities of Scotland with Evidence and Appendix*, iii (Edinburgh, 1878), *Parliamentary Papers* (1878) (C–1935), p. 287, as quoted by M. Dupree in 'The Development of Medical Education from 1870–1940', paper presented to the University of Glasgow 11th Jubilee History Seminar on Medical Education, 1 February 1996. I am grateful to the author for a printed copy of this paper.

[85] Gairdner, 'Memories of College Life', pp. 61–62.

[86] Ibid., p. 62.

as two-time President of the Faculty had also been a surgeon at the GRI (1846–50) and from 1850 one of its managers, commented in 1876 both the community and the hospital needed a medical school at the GRI:

> It was hard to see the weapons in our armoury rust; hard to behold the rich and fertile field for clinical instruction comparatively poorly reaped ... disease can only be profitably studied in an hospital, especially in an hospital of adequate dimensions. It is there that it presents itself in all its protean aspects; there it is studied and classified by competent observers. It is therefore a matter of vital importance not only to the medical profession, but to the whole community, that such institutions should be fully utilised for educational purposes. A medical school, clinical or systematic, or both, is essentially the natural complement of every large hospital. Even from the point of view of the efficiency of the hospital itself as a curative institution, it is most desirable that the medical officers should have a large following of students in the wards.[87]

Fleming argued that the scrutiny of students eager to learn improved the quality of patient care by making physicians and surgeons more thorough,[88] and, in any case, the hospital needed students as a source of labour: forty dressers alone were needed in surgery to keep the hospital running efficiently.[89] After initial plans for the amalgamation of the two infirmaries, and the running of the GWI as a branch of the GRI had quickly fallen through, the managers of the GRI therefore decided to found their own medical school on the model of the London teaching hospitals.[90] A supplementary charter was obtained in 1875 from the Queen in Council allowing teachers to be employed to instruct students, and the Royal Infirmary School of Medicine was opened in 1876. New buildings, including substantial laboratory accommodation (essential to compete for students with the University after the GMC had stressed the importance of basic sciences in the curriculum in the 1860s), were completed by 1882. The new medical school had lower fees than the university and students studied for the TQ, so this was an important development for the FPSG as well as for the GRI, a fact reflected in Fleming's instrumental role in the establishment of the new school. However the GRI School suffered a further blow in 1888 when Anderson's Medical College also moved west to a site near the University in Dumbarton Road, and many more students left the wards of the GRI for those of the GWI.

---

[87] John Gibson Fleming, 'Address Delivered at the Inauguration of the Glasgow Royal Infirmary School of Medicine on 1 November 1876' (Glasgow, 1876), pp. 10–11.
[88] Ibid., p. 11.
[89] 'The Glasgow Medical School after 1850', hand-written paper (possibly by Stanley Alstead), RCPSG, 1/13/7/8.
[90] Fleming, 'Address Delivered at the Inauguration', p. 11.

This was in turn countered by the establishment of a new college of medicine, St Mungo's, which took over the GRI Medical School from the infirmary's managers in 1888.

Given these kinds of rivalries it is hardly surprising that the idea of a conjoint scheme including the universities eventually came to nothing in Scotland. The GMC officially despaired of ever achieving one, given the implacable opposition of the Scottish universities. Instead, it decided to work for improvement in national examination system through the existing arrangements, and increase its visitations and reports.

Examinations and the General Medical Council

The GMC was given power to inspect examinations by visiting them under the 1858 Act. This did not begin immediately but between 1866–68, 1873–75, and again after 1881 a concerted campaign of visitations was conducted. In the 1870s, as noted, the campaign was prompted by a desire to achieve voluntary reform along the lines suggested by Simon and so to make obsolete the stalled medical legislation. Better for the profession to reform itself than to be reformed by a government which might take additional rights away, and leave the future government of the profession in the hands of a reformed GMC which included directly elected general practitioners. The GMC sent visitors to observe and report on the Faculty's licence in August 1873 and on the Double Qualification in January 1874 as part of its intensified campaign. The Council had, as has been noted, little power to coerce compliance, though the medical authorities were bound to be even more circumspect in falling into line in the post 1870 climate when competition between exam-inations was intense. Individual authorities, of course, would also have shared the GMC's perception that voluntary reform would preserve more rights and privileges.

While the market for medical expertise exerted pressure for comprehensive combined examinations, the GMC's redoubled efforts pushed ahead reform of single surgical qualifications for general practitioners, who didn't necessary need a formal double qualification if they did not intend to enter the public service. Such men would establish a private practice, and possibly seek a local hospital appointment. As Newman has argued,[91] one of the GMC's key contemporary objectives was to raise the standard of medical knowledge among general practitioners. This was the period when science came to play an increasingly important role first in diagnosis and then in treatment. Examinations had to be tightly monitored to ensure they kept up with GMC

---

[91] Newman, *Evolution of Medical Education*, chapter 5.

recommendations for broadening the standard medical curriculum to include more basic science and the use of medical technology.

Apart from public safety, the GMC also had in mind the public image of the profession. The great mass of the profession were humble GPs, some of whom had only recently entered the profession proper from the underclass of surgeon-apothecaries because of the necessity of qualification for registration after 1858. The medical elite felt, no doubt remembering the pressure for more democratic government of the profession from bodies such as the BMA during the 1840s, that such men, who were often of a lower social class, must be assimilated into the mores of the upper middle-class profession: they must learn to think of themselves as Victorian gentlemen for the cohesion of the profession. The appeal to science and scientific method was a characteristic strategy of advancement in a range of professions in the last third of the nineteenth century. Science was fashionable and commanded considerable social and cultural authority. The conflation of the concept of professional with that of scientist provided a new way not only of justifying but of increasing the power and prestige of the professions.[92] The GMC was underlining the social status of the profession to all of its members by increasing the scientific content of examinations.

The reports of the Visitors offer an insight into the nature and scope of the Faculty's examinations in this period which is difficult to find in any other source. The Faculty granted (and relied on fees for) many more ordinary surgical licences than it did double qualifications throughout this period. The 1873 visitation of the Faculty's licence covered the first and second examinations, omitting the final.[93] The first examination comprised anatomy, physiology and chemistry, and the second, surgery and surgical anatomy, practice of medicine, materia medica, midwifery and medical jurisprudence. Both examinations had written and oral components. In the first examination, unfortunately, on this occasion, the examiners in anatomy and physiology were both absent and the orals were conducted by Dr James Morton, one of the examiners in surgery. The *viva voce* examinations lasted twenty minutes in anatomy, and only five in physiology. In the anatomy examination a tray of 'various bones and a dried specimen of a dissected

---

[92] Much work has been done on this subject. A good place to start is the work of Frank M. Turner, and especially his two articles: 'The Victorian Conflict between Science and Religion: A Professional Dimension', *Isis*, 69 (1978), pp. 356–76; and 'Public Science in Britain, 1880–1919', ibid., 71 (1980), pp. 589–608. Both contain useful footnotes to guide further reading.

[93] See *GMC Minutes*, 11 (1874), pp. 90–105, 'Report of the Visitors Appointed to Inspect the First and Second Examinations of Candidates for the Diploma of the Faculty of Physicians and Surgeons of Glasgow', held at Glasgow, 12–13 August 1873.

lower extremity with the large vessels only injected, the arteries and veins coloured respectively red and blue' appeared on the table but none were utilised.[94]

While approving of the chemistry examination, the visitors were not so happy with those in anatomy and physiology, feeling that both were too cursory. They began their report opining that, since the Faculty's professional examination was divided into two parts to ensure the requisite weighting for important subjects, 'a high standard of knowledge' should be 'insisted on by the Examiners, since the student is able to give his undivided attention to each class of subjects long enough to acquire a thorough practical knowledge of them'.[95] The report later detailed specific criticisms:

> whilst the printed Anatomical Questions are perfectly fair and well adapted to test the acquirements of the Candidates, they only represent ... one branch of a thorough Examination in Anatomy; and they feel that the estimate of the Examiner would have been more satisfactory had the Written Examination been supplemented by a good *viva voce* Examination on the muscles, bones, and ligaments, and by insisting on each candidate dissecting some part or region of the body. It is clear that the object of the above-mentioned division of subjects for Examination is defeated if it be known that a student may pass, who, when asked so large a question as the distribution of the eighth pair of nerves, upon so important a subject as Anatomy, can say all he has to say upon it in eight printed lines. In ... regard to Physiology ... it is impossible for an Examiner to obtain from one paper of three questions (for which only an hour is allowed) and a *viva voce* of five minutes, such a knowledge of a Candidate's attainments as shall justify him in passing, though it may be quite sufficient to demonstrate the necessity of rejecting the Candidate.[96]

The Visitors also felt that all candidates for the first examination should demonstrate knowledge of the simple tissues under the microscope, but felt that all the shortcomings they had identified could possibly be explained by the absence of the two special examiners.

The second (or pass) examination fared a little better. The written part received no comment, while the twenty minute orals in surgery, surgical anatomy, practice of medicine, materia medica, medical jurisprudence and midwifery won general approval. The Visitors were impressed with the universal use of specimens and the depth of questioning, and reported that these examinations were, 'sufficiently extensive and varied to furnish a good test of the Candidate's attainments'.[97]

[94] Ibid., p. 94.
[95] Ibid., p. 98.
[96] Ibid., pp. 98–99
[97] Ibid., p. 104.

The Visitors had, however, reservations about the conduct of the clinical examinations. As was usual, a clinical examination in both medicine and surgery took place at the Royal Infirmary, on this occasion directed by Robert Scott Orr and Robert Perry (medicine), and James Morton and Donald Dewar (surgery). In medicine the candidates (examined separately) were directed to examine, diagnose and prescribe treatment for the same case of Bright's Disease, and then one of pleurisy, and of jaundice, respectively. Urine specimens were then introduced and the candidates asked how to test for albumin, and about interpreting urinary evidence of disease. Urinary deposits were examined microscopically and the candidates were questioned on the findings. In the clinical surgery examination the candidates were taken into the surgical wards and required to examine, defend a diagnosis on, and prescribe treatment for three cases: a tumour in the right iliac region; a fractured knee-cap; and an apparent severe fracture of the left tibia. On this occasion only one pass, in clinical surgery, was awarded out of two candidates.[98]

The Visitors felt, again, that the examination in clinical medicine was not long enough. They suggested that written reports on two or more cases would be a useful measure of a candidate's ability, since it was difficult for the examiner to avoid leading questions when he conducted the examination at the bedside with the candidate. They also felt that more microscopical and pathological specimens should be exhibited. In the longer surgical examination the Visitors lamented the lack of instruments used; the lack of practical examinations in bandaging, and the management of fractures and dislocations; the lack of any display of pathological specimens; and the lack of any test of practical surgery on a cadaver. The Visitors argued that the Faculty should give these points 'careful attention'. Their use gave important insights into the candidate's knowledge and the Faculty's regulations specifically stated that recent dissections, anatomical specimens, surgical apparatus and pathological specimens could be brought in at the examiners' discretion. In fact, in the anatomy oral and in the examination in clinical surgery, such aids had not been utilised.[99]

In January 1874 the Visitors attended the second examination of the Double Qualification. This time, apart from minor technical criticisms of the conduct of the examinations, they were generally satisfied that they insured that 'those who receive the Conjoint Diploma of the Glasgow Faculty, and the Edinburgh College are competent to practise their profession'.[100] In the controversial

[98] Ibid., pp. 102–3.

[99] Ibid., p. 104

[100] 'Report on the Second or Pass Conjoint Examination of the Faculty of Physicians and Surgeons of Glasgow, and of the Royal College of Physicians of Edinburgh', GMC Minutes, 11 (1875), pp. 80–89.

clinical part the Visitors noted with obvious approval that they were more rigorously conducted than previously. In medicine, the candidate was quizzed on urine samples:

> He was then sent to examine a patient, and after a certain time, when he was supposed to have mastered the case, the Examiners went to the bedside and called upon him to explain the nature of the disease and to show how he had examined it. A second case was then given to him, and the same method was pursued.[101]

In the surgical part, as well as the cases, there were tests of bandage and splint use in fractures, and of the use of the tourniquet. The Visitors believed that the bedside examination was now a much more stringent test of whether the candidate had learnt the art of clinical diagnosis:

> It was a fair test and while no properly prepared man could have failed in it, an ignorant man would certainly be detected. The cases of disease both Medical and Surgical were well selected, and there was sufficient testing of the application of Surgical apparatus.[102]

The examinations, however, had been carried out with only two candidates and the Visitors 'hardly know how the Examiners could have carried out the Clinical Examination on the plan followed, had there been ten or twenty Candidates instead of two'.[103]

In accordance with GMC procedure, the Faculty commented on the Visitors' report. Most of the criticisms were reluctantly accepted and only the remarks on the clinical medicine examination were objected to. The Faculty argued that no other examination for a surgical licence demanded a clinical examination in medicine, and that this should be taken into account when judging this part.[104]

The next set of GMC visitations occurred during 1881–82, and the reports were then even more positive. In the DQ and the single licence, the inspectors praised the high quality of the recently introduced practical anatomy examination, adjudging it to be 'in advance of the London and Edinburgh Colleges'.[105] They also admired the clinical examinations in medicine and surgery conducted at both the Royal and Western Infirmaries, which included

---

[101] Ibid., p. 82.

[102] Ibid., p. 83.

[103] Ibid.

[104] 'First and Second Examinations of the Faculty of Physicians and Surgeons of Glasgow', *GMC Minutes*, 12 (1876), 25 June 1875, pp. 108–9.

[105] 'Evidence and Reports on the Examination of the Faculty of Physicians and Surgeons of Glasgow', by 'J. F.', *GMJ*, 18, series 5 (1882), pp. 258–86. Quotation is taken from, 'Extracts from the Report on Visitation of Examinations, 1881–82', reproduced herein, p. 272.

microscopical examination of urine and tests of bandaging and the use of splints. The general comment of the inspectors was that 'the aim and plan of the examinations of this body are essentially good, and are evidently improving on correct lines'.[106]

The Triple Qualification

While piecemeal advances could be achieved by the GMC with its existing limited powers, there was still a majority in the GMC committed to wholesale centralised reform. There were also various other questions which needed to be cleared up (for instance the status of foreign qualifications and the admission of women to examinations). In May 1877, a GMC deputation pressed the Lord President, now the Duke of Richmond and Gordon,[107] for a government Bill covering these points, and, as in 1870, he replied that if the government took up medical reform it must address the whole issue. A new comprehensive government Bill was thus introduced on 19 March 1878. It required a double qualification for registration: either two diplomas (one medical and one surgical) from medical authorities recognised in Schedule A of the 1858 Act, or a qualifying certificate from a new joint board and a single diploma from a medical authority. Candidates who passed the joint board examination were entitled to the qualification of any one body which participated in the board, so that the Bill made affiliation to a recognised medical authority mandatory. However, there was no requirement on medical authorities to combine into joint boards.[108]

The Faculty petitioned in favour of this Bill stating that it offered a 'fair and proper' solution to long-standing difficulties. Armed with its double qualification, the Faculty had nothing to fear from such a Bill: 'The Bill is a judicious and well considered measure and that if passed into law it would, without introducing violent or radical changes effectively contribute to bringing about a reform which would at once tend to elevate the medical profession and benefit the public'.[109]

In England, however, the RCPL, the RCSEng, the Apothecaries' Company and the University of London objected to the voluntary conjunction in the Bill. They had signed up to a joint board scheme in 1877 (though this was not yet operative), and they feared that unless such combination was obligatory, they would lose business to easier (and they meant Scottish)

---

[106] Ibid.
[107] Appointed Lord President in 1874 and Conservative leader in the Lords.
[108] The Bill also dealt with foreign degrees, women, dentists, midwives and aspects of lunacy.
[109] Faculty Petition, *Faculty Minutes*, 6 May 1878.

non-combined qualifications. In April the GMC narrowly resolved once more in favour of compulsory conjunction by fourteen votes to ten. All the Scottish and Irish delegates voted against, and the University of Glasgow presented a petition to the House of Lords expressing satisfaction that compulsory conjunction formed no part of the Bill. Having failed to agree on a joint board scheme, the Scottish medical authorities now all supported the Scottish universities' position against conjunction. The Scottish authorities were content with the existing situation, but in terms of public rhetoric at least the issue had become a national one: the protection of the special position of the Scottish universities versus the unfair levelling schemes of England. A single portal would weaken the stimulus to higher qualifications and thus lower the standard of medical education. This would particularly affect the Scottish universities.

At the Bill's second reading on 15 March Lord Ripon, the previous Lord President who was still working on Simon's agenda, managed to get the Bill amended to resemble the more radical 1870 Bill. Simon, President of the RCSEd and crown member of the GMC since May 1876, was again whipping up support for the single portal. Conjunction was again made compulsory, and affiliation was dropped: now the qualification of a conjoint board alone was necessary for registration. This completely undermined the position of individual medical authorities and was universally rejected in Scotland. In petitioning against the amended Bill, the Faculty stressed that abolishing affiliation and introducing compulsory conjunction not only threatened the existence of medical authorities by making their qualifications unregistrable, but also destroyed the delicate balance of the organic British system of professional regulation. The medical authorities (and the Faculty meant corporations like itself) were not just examining bodies but professional organisations which superintended and disciplined members. Bodies like the Faculty, with their medical libraries, scholarships and annual scientific lectures, provided a focal point for the local profession, binding it together not only with a shared professional ethos but also with a shared interest in the advancement of medical science. The Faculty felt that the amended measure was too revolutionary and asked that the whole matter of medical licensing be referred to a Royal Commission or Select Committee.[110]

The GMC formally objected to the dropping of affiliation and the government, anticipating trouble in getting the amended measure through the Commons dropped the Bill in July 1878. It was reintroduced on 25 February

[110] See *Faculty Minutes*, 1 July 1878, Report of Council and Faculty Petition against amended 1878 Bill.

1879,[111] with amendments meant to placate the profession but was referred by the government to a Select Committee, mainly because of opposition from the BMA about direct representation of the profession on the GMC, which the Bill did not tackle. Again the Scottish medical authorities had opposed the Bill anyway. The Faculty had again petitioned the Lords arguing against compulsory conjunction on the grounds that competition between medical authorities in Scotland continually elevated the standard of examinations, and that the visitations of the GMC were the most effective way to ensure further improvement. The Faculty pointed out that since it was anticipated that each conjoint board should draw up its own regulations, 'uniformity of examination (except in so far as already obtained by uniformity in the subjects for examination and the mode of conducting it) is a chimerical object which no medical Bill can ever accomplish'.[112]

In May 1881 a Royal Commission was appointed to continue consideration of the medical reform question.[113] The main recommendations of its 1882 report were that a double qualification should be required for registration, that conjoint divisional boards should be established in each division of the kingdom to administer the examinations, and that the GMC should be reduced from twenty-three to eighteen (four members elected directly by the profession; eight elected by the divisional boards; and six crown nominees). A new government Bill was introduced into the Lords in March 1883 based on the commission's findings. There was to be no registration without qualification in medicine, surgery and midwifery; Medical Boards were to be established in each division to regulate examinations, under GMC supervision; women were to be admitted equally with men; and the GMC was to be reformed.

The Bill was received favourably by the profession. Compulsory conjunction and a single portal now seemed such a *fait accompli* that there was little further public opposition, although all of the Scottish witnesses before the Select Committee and Royal Commission had been against such a plan. The Faculty had finally accepted that there should be a single portal in Scotland and restricted its opposition to the Bill to complaining about the representation on the new divisional Medical Board. In Scotland the board was

[111] The Bill was also reintroduced in 1880 and re-referred, together with various private Member's Bills, to the Select Committee which continued to sit during 1879–80. See *Parliamentary Papers* 320 (1879) and 121 (1880).

[112] *Faculty Minutes*, 7 April 1879, Faculty Petition against government Bill of 25 February 1879. The Faculty sent largely the same petition to the House of Commons, *Faculty Minutes*, 5 May 1879.

[113] 'Report of the Royal Commissioners Appointed to Inquire into the Medical Acts', *Parliamentary Papers* (1882) C. –3259. L.

to be composed of three representatives from the university of Edinburgh; two from Glasgow; two from Aberdeen; one from St Andrew's; and one each from the Scottish corporations.[114] The Faculty argued, borrowing the BMA's successful argument for direct representation of the profession on the GMC, that three quarters of the Scottish board would be appointed by elite university senates, most of whom were not even members of the profession. This would leave the democratically elected corporate representatives in a powerless minority, and would thus threaten the existence of the extra-mural schools of medicine. The Faculty argued that the 'general body of the profession' should be 'directly represented on these Boards in the way the Bill provides for their direct representation on the GMC'.[115] Meetings were held between the Faculty and the Edinburgh Colleges in April and May 1883 and it was agreed that the corporations should press for nine representatives on the Scottish Medical Board (the universities retaining eight), and that one of the two GMC representatives from the board should be nominated by the corporate representatives.[116] In Parliament the Bill was fatally delayed by Members acting for the Scottish and Irish corporations trying to secure such changes to the structure of the boards, and, once again, the Medical Bill was withdrawn.[117]

There followed a flurry of activity in both England and Scotland as the medical authorities again tried to circumvent the need for radical reforming legislation with limited voluntary action. The English Conjoint Board Scheme was finally got up and running in October 1884. In Scotland negotiations between the three corporations to restrict the granting of independent single licences and instead to institute a combined Triple Qualification (TQ) of the *three* bodies (in medicine, surgery and midwifery) had begun in November 1883.[118] The scheme was created to establish what the Faculty referred to as a 'complete licence', so that compulsory conjunction into a joint board including the universities in any future legislation, could be opposed on the basis that a comprehensive conjoint corporate examination already existed. This would leave both corporations and universities as relatively independent as they could be, given the reforming trend of imminent legislation. The Faculty Council was now prepared to describe the single licence as a 'half-qualification' and recommended that:

---

[114] See 'The Medical Act Amendment Bill, 1883', as reprinted in the *British Medical Journal*, 1 (1883), pp. 578–84.

[115] *Faculty Minutes*, 17 April 1883, Faculty Petition against 1883 Bill.

[116] See *Faculty Minutes*, 7 May 1883.

[117] See Cooke, *History of RCPL*, p. 860.

[118] See *Faculty Minutes*, November 1883, passim.

The time had now come when the granting of half qualifications or single licences should cease and that [the three Scottish Corporations] should come to an Agreement that such licences should be granted by them only in cases in which the candidate possessed another and apposite qualification.[119]

It was not just preempting the imposition of a conjoint board including the universities which had led to such a critical attitude towards the licence (which, as we noted, had a curriculum almost as broad as the DQ). Rather, during 1881–82 Scottish qualifications in general, and those of the Faculty in particular, had come under sustained attack and the Faculty had been accused of offering substandard examinations. Once again the source was English.

In 1881 the *Glasgow Medical Journal* reported in an editorial that the Faculty was 'in serious danger of being snuffed out, so far as its public functions are concerned'.[120] The Medical Reform Committee of the British Medical Association had adopted a resolution lumping the Faculty together with the Apothecaries' Society of London and the Apothecaries' Hall in Ireland, and calling for all three to 'cease to exist as medical authorities'.[121] This resolution was passed at a conference consisting of delegates from four bodies: the Medical Reform Committee; the Medical Alliance Association; the proprietors of the *Lancet*; and the Irish Medical Association. As the writer, the editor of the *GMJ*, Joseph Coats, then pathologist to the Western Infirmary and, though not a professor until 1893, responsible for all the pathological teaching of university students at the GWI,[122] and an active Fellow and Council member, dryly pointed out, the first consisted of sixteen English practitioners and a member of the Irish clergy; the second used to be called the East London Medical Defence Association, was formed to campaign against quackery and for the 1858 Act and, as such, 'can hardly be regarded as representative, except of a group of London practitioners'; the authority of the third, a 'valuable commercial concern', to legislate for the medical profession was non-existent, and again London orientated; and of the fourth it could only be said that it, 'may or may not be representative of the Sister Isle, but certainly does not represent Scotland'.[123] When the resolutions that came out of this conference were adopted by the BMA, again there was no Scottish presence, perhaps this time because, in time-honoured fashion, it was presented at the end of the meeting when most Scottish delegates would

[119] *Faculty Minutes*, 21 December 1883, communicating Council Resolutions of 18 December.
[120] 'Current Topics: The Faculty and the Medical Reform Committee of the British Medical Association', p. 359, *GMJ*, series 5, 16 (1881), pp. 359–63.
[121] Ibid., p. 360, quoting the Medical Reform Committee's Resolutions.
[122] See obituary of Coats by W. T. Gairdner, *BMJ*, 4 February 1899.
[123] All quotations in this section are from *GMJ*, 'The Faculty and the Medical Reform Committee', pp. 360–61.

have already begun the long train journey north. Coats attacked the resolution calling for the end of the Faculty as a medical authority, citing in its defence the recent favourable GMC reports on its examinations conducted during 1881–82, which we noted above:

> We could understand a comprehensive scheme by which the authorities were consolidated and united, and their licences given on some common principle, but this process of tinkering at the matter is hardly worthy of a body calling itself the British Medical Association. The fact is not to be concealed that the Faculty licence has a bad name in England, and we have no doubt that the Glasgow School of Medicine, as a whole, suffers by reason of that. But there is evidence in recent results that the examinations are now satisfactory, and that, as a medical authority, the Faculty is able to compare with the other authorities in the Kingdom.[124]

Coats felt that, 'this whole outcry about Medical Reform has intimate relations to the great success of the Scottish Schools of Medicine'.[125]

The controversy over the status of Scottish qualifications in general, and those of the Faculty in particular, continued into 1882 with the publication of the report of the Royal Commission on the Medical Acts. The evidence to the commission contained many, usually second-hand, critical observations on the Faculty's examinations. All of the reservations were expressed by English medical witnesses and most mentioned that the Faculty's examinations were easier, covered fewer subjects, and were used as a last resort for those failing London qualifications.[126] Defending the Faculty's examinations in the *GMJ*, 'J.F.', a hospital practitioner who took part in the Faculty's clinical examinations, but had no other official connection,[127] noted that while the Faculty's single licence was technically a surgical qualification, yet:

> In common with its sister College in Edinburgh, the Glasgow Faculty has always recognised the necessity of considering the candidate's ability for general as well as purely surgical practice. In the charter of the Faculty, procured in 1599 from James VI, the founder, Peter Lowe, although himself a surgeon and the author of the earliest British treatise on surgery, secured the cooperation of a physician,

---

124 Ibid., p. 362.

125 Ibid., p. 363.

126 For examples of the adverse criticism, see 'Evidence and Reports on the Examination of the Faculty of Physicians and Surgeons of Glasgow', *GMJ*, 18 (1882), pp. 258–86.

127 'J. F', may not be initials but may indicate a member of the GWI staff. In 1878 the GWI received a bequest of £40,000 from John Freeland from his family's trading firm, which allowed completion of the original hospital plan, and thus a doubling of the existing ward accommodation. The monogram 'J.F.' adorned the three wings of the extension. See Loudon MacQueen and Archibald B. Kerr, *The Western Infirmary, 1874–1974* (Glasgow, 1974), pp. 11–12.

and even of an apothecary, for the due discharge of the important duties of this incorporation ... True to its traditions the Faculty has endeavoured to determine that its licentiates were fit for general practice.[128]

The author also emphasised that the Faculty's first examination for its single licence covered more subjects than the equivalent exam of the English surgeons, had examined in physiology long before that body, and included chemistry, a subject still not included in the RCSEd examinations. The Faculty examination had also long included materia medica, medical juris- prudence and midwifery, all essential subjects for the would-be general practitioner. Also included was an examination in clinical medicine, in spite of the fact that this was known as a surgical licence.[129]

This kind of sustained criticism, as well as the desire to preempt more sweeping national reforms, made the Faculty eager to enter into arrange- ments to link its qualifications with those of *both* Edinburgh Colleges in a conjoint board to shore up the status of its exams in the light of their poor professional image, however unwarranted. The arrangements for a Triple Qualification (TQ) to supersede the two DQs were approved by the Scottish corporations and sanctioned by the GMC in March 1884.[130] All parties agreed not to award their own single diplomas independently other than to candi- dates with a complementary qualification (recipients of the Physicians' diploma must have a surgical qualification from a Schedule A body, and of the Surgeons' and the Faculty's diplomas, a diploma from a British College of Physicians). The new TQ examination retained the old tripartite structure. The first examination now covered chemistry, elementary anatomy and histology; the second, anatomy, physiology. materia medica and pharmacy; and the third, principles and practice of medicine (including therapeutics, medical anatomy and pathology), clinical medicine, principles and practice of surgery (including operative surgery, surgical anatomy, and pathology), clinical surgery, midwifery and diseases of women (including operative gy- naecology, medical jurisprudence and hygiene). The examinations were to be held in both Glasgow and Edinburgh. In Glasgow a joint board of exam- iners from each party would examine candidates on each subject, the composition of the board varying depending on whether the subject was medical or surgical. The fee was twenty-five guineas, of which two tenths went to the Faculty, three to the Edinburgh Surgeons and the rest to the

---

[128] *GMJ*, 'Evidence and Reports on the Examination of the Faculty of Physicians and Surgeons of Glasgow', p. 260.

[129] Ibid., p. 261.

[130] Craig, *History RCPEd*, pp. 308–11.

Physicians. On passing the third examination the candidate was entitled to style himself LRCPEd, LRCSEd, and LFPSG.[131]

The poor image that Faculty's examinations had within the profession which had been revealed during 1881–82, also spurred the Faculty to reform the Fellowship at the same time as discussions were going on to implement the TQ. In October 1883, after a couple of thwarted attempts, Mr John C. McVail, a physician at the Kilmarnock Infirmary, MOH for the counties of Stirling and Dumbarton, and staunch supporter of vaccination,[132] finally got a motion to award the Fellowship only after an examination passed, after a split vote, decided on the vote of the President, Dr Andrew Fergus.[133] Examination had been the original method of awarding the Membership, as it was called until the Faculty Act changed Members into Fellows, but it had changed into an honorary title conferred only after an election by other Fellows sometime in the early nineteenth century.[134] The Faculty was now acutely sensitive to accusations that it was not a *bona fide* medical authority and an examined Fellowship would underline its status as a professional scientific body and not just an elite local club. That this reform was linked to the reform of the DQ and single licence into the TQ is underlined by the fact that the other motion that McVail managed to get through at the same meeting was to open discussion with the Edinburgh Colleges to award 'a complete licence in Medicine, Surgery and Midwifery'.[135] A further motivation to reform the Fellowship was that the Edinburgh Surgeons were also moving to reform their Fellowship to bring it into line with the examined Fellowship/Membership of the London Colleges.[136] In fact the President had received a letter from the Edinburgh Surgeons telling him this on 5 November 1883. Clearly the Faculty could not be seen to be behind other Scottish medical corporations in the prevailing climate of professional hostility to all Scottish qualifications.

In February 1884 the Faculty backed the Council's general recommendation that: 'the Fellowship Examinations to be instituted be of a thoroughly testing

---

[131] Details of TQ from 'Scheme of Conjoint Examination to be Conducted by the Royal College of Physicians of Edinburgh, the Royal College of Surgeons of Edinburgh and the Faculty of Physicians and Surgeons of Glasgow', *Faculty Minutes*, 3 March 1883.

[132] See *Medical Register*, 1912.

[133] *Faculty Minutes*, 15 October 1883.

[134] For more details on this see Johanna Geyer-Kordesch and Fiona MacDonald, *Physicians and Surgeons in Glasgow: The History of the Royal College of Physicians and Surgeons of Glasgow*, i, *1599–1858* (London, 1999).

[135] *Faculty Minutes*, 15 October 1883.

[136] See C. H. Cresswell, *The Royal College of Surgeons of Edinburgh: Historical Notes from 1505–1905* (Edinburgh, 1926), pp. 296–98.

character, the standard begun and maintained as high as that of any of the existing Fellowship Examinations'.[137] The actual content of the examination however proved to be more difficult to agree, and discussions within Faculty dragged on until February 1885,[138] when the scheme of examination was finally settled. In this first version of the modern Fellowship of the Faculty there were two basic groups of subjects. The first contained subjects covered in the licence, and exemptions were given to Licentiates; the second contained subjects unique to the Fellowship which were to be examined at a higher level. Group one consisted of medicine or surgery (surgery for those selecting medicine in group two and vice-versa), chemistry, materia medica, midwifery, and medical jurisprudence. Group two consisted of surgery or medicine, anatomy, physiology, and pathology. These subjects were covered in two separate examinations. The first covered anatomy, physiology and chemistry. Dr William Macewen, the internationally famous Regius Professor of Surgery at the University of Glasgow, a neurosurgical pioneer with a keen interest in the latest scientific techniques and discoveries, made sure that this examination was searching and up-to-date. While it was reviewing the arrangements for the Fellowship, the Faculty approved his motion that:

> The examination in Anatomy shall comprise oral questions on recent dissections, preparations and models. The candidate will be expected to show a thorough acquaintance with the Anatomy of the brain and organs of sense, with the development of the different organs and the process of ossification. A knowledge of the best modes of preparing and displaying the different regions of the body will also be expected ... For the examination in Histology [which, with Embryology, was common to Anatomy and Physiology] the candidate will be expected to have a satisfactory knowledge of the best methods of preparing and mounting microscopic specimens of normal tissues and will be required to prepare such specimens in the course of the examination. They will also be required to identify mounted specimens under the microscope. The examination in Embryology shall include the development of the ovum up to and including the formation of the several layers of the blastoderm. It shall also include the development of the brain and spinal cord and the circulatory and osseous systems. In Physiology proper the candidate shall be required to show a sound knowledge of the subject and also a practical acquaintance with the apparatus employed in physiological physics and a knowledge of the practical chemical examination of blood, urine, milk and bile.[139]

The subjects of the second examination were medicine or surgery, pathology, materia medica, midwifery and medical jurisprudence. Again the standards were high. In pathology the candidate would be quizzed on both

---

[137] *Faculty Minutes*, 4 February 1884, RCPSG, 1/1/1/11.
[138] Ibid.
[139] *Faculty Minutes*, 10 November 1884.

macro- and microscopic specimens; if surgery was taken the candidate would be examined in systematic surgery, including surgical anatomy and surgical pathology; in operative surgery, including the performance of operations on the dead body and knowledge of surgical instruments; and in clinical surgery involving the examination of patients and the use and application of surgical apparatus. If medicine was taken, the candidate would be examined in systematic medicine, including medical pathology and therapeutics, and in clinical medicine, again involving the examination of patients.[140] The fee for the new examination was to be fifteen guineas to existing Licentiates or University graduates and twenty guineas to others. Interestingly the Fellowship was initially aimed at young practitioners who had not long finished their initial medical studies, as an indicator of their potential to go on to a career in the higher reaches of medicine. However, in 1884–85 it was felt that this conception excluded more experienced practitioners who could not afford the time away from their hospital or private practice to return to study. So a special examination was set up for those practitioners who had been qualified for at least ten years. They were tested in either medicine or surgery and one subject from human anatomy, comparative anatomy, physiology, pathology, midwifery and the diseases of women, medical jurisprudence, state medicine, and psychological medicine. Provision was also made for the admission to the Fellowship of not more than two practitioners a year who were specially distinguished in medical science, usually measured by publication, and at the discretion of the Council. Thus the first Fellowship exam of modern times was not directly modelled on the new Fellowship examination of the RCSEd, but was rather more comprehensive.

The new regulations for the examined Fellowship came into force in 1885 and the first candidate to be admitted a Fellow under the new system was Dr T. K. Dalziel in 1887.[141] The examined Fellowship became registrable as an additional qualification with the GMC on the Medical Register, making it a more attractive qualification, and one that the Faculty hoped would become an essential prerequisite for local hospital appointments. This new qualification which began in such a small way in 1885 was to become the rock on which the Faculty built a new future, after postgraduate education

[140] Ibid.
[141] Sir Thomas Kennedy Dalziel, MB, CM Edinburgh 1883. He was Surgeon to the Royal Hospital for Sick Children and the GWI in Glasgow. He was Lecturer on Medical Jurisprudence, then Professor of Surgery at Anderson's College. He worked and published in gastroenterology, being one of the first to identify what later became known as Crohn's Disease. He was also a medical referee under the Workman's Compensation Act and a member of the Edinburgh Royal Medical Society and the Glasgow Pathological and Clinical Society. He died on 10 February 1924. Details from Biographical Index of Fellows, RCPSG.

had become crucial to the full-time hospital doctors of the National Health Service.

Meanwhile, in March 1884, the previous year's government Bill was reintroduced with minor amendments, one giving two more representatives on the Scottish Medical Board to the corporations (one for the RCPEd and one to alternate between the Faculty and the RCSEd). The Faculty still did not feel this was enough and said so in a petition to the House of Lords. The Faculty also objected to a clause allowing university students to be examined at a lower fee, again fearing that this would wreck the extra-mural system with which (via Anderson's and the GRI Medical School) it was intimately linked and which operated as its teaching arm. However, the main objection was that legislation enacting compulsory joint boards was now unnecessary in Scotland as the TQ had superseded it. As the Faculty argued, the TQ 'will effect in a manner much simpler and less expensive than that provided for in the Bill the same results viz. conferring complete qualifications to practise, after an efficient examination conducted by a United Board'.[142]

The other notable aspect of the TQ was that it was open to women students. This was in response to the passing of the 1876 'Act to Remove Restrictions on the Granting of Qualifications for Registration under the Medical Act on the Ground of Sex'. The first recorded female pass in the First Examination was that of Grace C. C. MacKinnon of Lerwick on July 8 1886, but first to pass the Final Examination and thus gain the TQ were Janet Hunter of Ayr and Sarah Gray of London who both passed in July 1888.[143] However the Fellowship was still not open to women. The Faculty had acted however on this controversial issue in advance of the University of Glasgow who only formally admitted women students after the 1889 Universities (Scotland) Act. The first women to graduate in medicine from any Scottish University got their degrees at Glasgow in July 1894 (Marion Gilchrist and Alice Louisa Cumming). These were students at Queen Margaret College. The College had established a medical school in response to student requests in 1890. The accessibility of the TQ to women, the only registrable Scottish qualification then open to them, was a stimulus for the foundation of the new medical school. The school was one of only four (with London, Dublin and Edinburgh) which at that time admitted women to medical studies.[144]

---

[142] *Faculty Minutes,* 7 April 1884, Joint Petition by the three Scottish Corporations Petition against Medical Bill.

[143] Details from RCPSG, 1/13/6/12.

[144] See Wendy Alexander, *First Ladies of Medicine: The Origins, Education and Destination of Early Women Medical Graduates of Glasgow University* (Glasgow, 1987); see also Johanna Geyer-Kordesch and Rona Ferguson, *Blue Stockings, Black Gowns, White Coats: A Brief History of Women Entering Higher Education and the Medical Profession in Scotland* (Glasgow, 1994).

As well as the establishment of the TQ and the reform of the Fellowship into an examined qualification, there had also been another notable development in the Faculty in these years. In 1878 the Dentists Act was passed regulating the qualifications of dentists, and thus both formally recognising them as part of the wider medical profession, and subjecting them to the same kind of national regulation. The RCSEd and the FPSG offered for the first time their own separate licentiateships in dental surgery. On 19 January 1879 the Faculty elected a Dental Board to examine in dental surgery, and the first candidates appeared for the Licence in Dental Surgery in April 1879. William Stead Woodburn of Glasgow was the first to sign the Glasgow LDS register. He subsequently became Lecturer in Dental Mechanics and Metallurgy at Anderson's Medical College, where the Glasgow School of Dentistry was formed in June 1879. The first women to pass was Williemina Simmons in 1901. The LDS remained the only undergraduate dental qualification until it was replaced by the BDS in 1948 when the University took over the Dental School.[145]

The interminable struggle for medical reform in this period was finally ended with the successful passage of the Medical Act Amendment Act on 25 June 1886. The Act followed William Turner's Minority Report from the 1882 Royal Commission. It was introduced by the vice-president of the Privy Council, Sir Lyon Playfair, who, tellingly, was the member for the universities of Edinburgh and St Andrews. The Act buried the idea of the single portal until it reemerged with the university monopoly of undergraduate teaching and examination after 1948 with the absorption of the extra-mural schools. While it prohibited the registration of any person not qualified in medicine, surgery and midwifery, there was no compulsory conjunction into joint boards (registrable qualifications could be awarded by universities or combinations of universities or corporations) and the GMC still had to report to the Privy Council to have unsatisfactory examinations changed. The whole saga of renewed agitation for reform in the years after 1858 had merely confirmed the entrenched power of the medical authorities. The idea of limited voluntary reform had worked in staving off more punitive reform. The Faculty, however, would have found a way of surviving relatively intact under what ever system had been established.

---

[145] Details here are taken from Carol Parry, 'Archives and Artefacts Relating to the History of Dentistry at the Royal College of Physicians and Surgeons of Glasgow', *History of Dentistry Research Group Newsletter*, 2, April 1998, pp. 14–16. Carol Parry is the College's Archivist.

The Faculty in Decline, 1886–1930

After the Medical Act of 1858 the Faculty of Physicians and Surgeons appears to have withdrawn from major disputations, but only the staunchest of supporters could claim that it made any great contribution to medical art or science. It is true that the library, tendered and fostered by a succession of devoted servants, became a repository for rare medical books and a place of recourse for scholars, and the monthly meetings became the forum for discussion on matters of everyday importance to medical men in the West of Scotland. But in the days of Lister, Gairdner, Macewen and their fellows it must be confessed that the fame and glory of medicine went increasingly westwards with the University.[1]

In the late nineteenth and early twentieth centuries the Faculty was at the zenith of its existence as an undergraduate licensing body offering alternative first entry qualifications in medicine to the universities. Most of the students at the Glasgow extra-mural schools studied for the TQ or the Licence in Dental Surgery (LDS – which particularly thrived during this period) for general medical or dental practice. However, once again national developments were germinating which would threaten what became a cosy existence. During this period the state, in both national and local manifestations, gradually took on increasing responsibility for the nation's health. The major landmarks were, of course, the introduction of National Health Insurance in 1911, the establishment of the Ministry of Health in 1919, and the long national debate on the form that a national medical service should take, which moved from the Dawson (1920) and Cathcart (1936) Reports with their ideas of a two-level system of primary and hospital care, to the final hospital/consultant based National Health Service in 1948.[2] But this brief list does not convey how in this whole period British political culture was infused

---

[1] Charles Illingworth, *Royal College of Physicians and Surgeons of Glasgow: A Short History and Tour Guide* (Glasgow, 1963), pp. 10–11.

[2] On these developments see, for example, Frank Honigsbaum, *The Division in British Medicine* (London, 1979); and David Hamilton, *The Healers: A History of Medicine in Scotland* (Edinburgh, 1981).

with concerns about the nation's health and exactly how to ensure it. This long and changing debate had a dramatic impact on the Faculty's social and professional role as an examining body guarding the standards for admission to medical practice, and finally threatened its very existence.

During this period the Faculty began to move towards a wider interpretation of its social role as a body representing the elite medical opinion of the city. After a slow start, it became more involved in giving its opinion, invited and uninvited, on national and local initiatives which affected the status of the medical practitioner. But it also became similarly involved, though only gradually and to a limited extent, in debates about the health and environment of the inhabitants of Glasgow. The Faculty continued to perceive itself chiefly as an examining body which oversaw the standard of entry to the profession.

By the end of this period it had, however, signally failed to appreciate the magnitude of the change that was imminent in medical education and practice. The reorganisation of health care provision envisaged during the 1920s and 1930s was largely debated in terms of an extension of National Health Insurance funded general practitioner cover to the whole of the population, with little change to the hospitals. This meant little incentive for a reform of the Fellowship, the qualification for consultants. The idea of a hospital and consultant based service came later, and implied even more dramatic changes in (postgraduate) medical education to train the new consultant elite. There were other trends, nonetheless, which might have but didn't force a fresh examination of the Faculty's core educational role in this period. The influence of the universities in undergraduate education gradually increased throughout this period, and the necessity for continuing, postgraduate, education of hospital consultants in an ever advancing and specialising medicine became ever more apparent. The Faculty failed to predict the impact that these trends would necessarily have on its role as an undergraduate examining body. Eventually, the Faculty was left high and dry with no professional role, since it had long neglected the status of its own higher postgraduate qualification, the Fellowship. As Tom Gibson has written,

> During the first forty years or so of the twentieth century, the Faculty's influence in medical training slowly decreased ... the post-[First World] war years with their economic depression and industrial and social unrest did little to encourage even an inclination for expansion and forward thinking in the Royal Faculty's affairs ... It became known locally as the 'Chum Club'[3] and nationally was slightingly referred to as the 'Photographic Society' from its initials. The main

[3] This from its oldest motto, *Conjurat Amice.*

incentive to becoming a Fellow seemed to be the possibility of becoming an examiner and earning some fees from the few students still presenting themselves for the 'Triple' or the Fellowship. In these days before the salaried Health Service, every little counted.[4] By the late 1930s, the Fellowship of the Faculty was regarded, even in the Glasgow area, as barely adequate qualification for a hospital appointment, and young graduates wishing to specialise took either an MD or ChM degree, or went to Edinburgh or London for a Royal College Membership or Fellowship; outside of Glasgow such a degree was a *sine qua non* for a hospital appointment.[5]

Another previous historian of the Faculty, Stanley Alstead, shared a gloomy view of this period, and pointed to disloyal members of the profession in Glasgow who turned their backs on the Faculty as partly responsible for the institution's increasingly moribund condition.[6]

All this was far off in 1885–86. At this point the Faculty seemed to be thriving. Income from both the TQ and the LDS was flowing in at a satisfactory rate. In 1885 the last year of the DQ, it had brought in £460, while the single licence yielded £396 28s. 0d. The new TQ yielded £700 in 1887 which had increased to £760 by 1888 while in the same years the revenue from the single licence had dropped from £182 14s. 0d. to a mere £19 19s. 0d., while by 1899 it had sunk to zero. The other success was the LDS. This brought in just over £136 in 1899 which had risen to just over £785 by 1910. The success of this examination temporarily disguised the waning of the TQ, which dropped to £400 in 1910 and maintained this low level for about the next twenty years.[7] Moreover, the future of extra-mural teaching at the Royal Infirmary, on which, as we noted above, the Faculty depended as a source of candidates for the TQ, was assured in 1892 when the Royal Infirmary School of Medicine was incorporated into the new St Mungo's School of Medicine.

In contrast, the status of the Faculty's Diploma in Public Health, aimed at

---

[4] Doctors were then obliged to compile a portfolio of small jobs in order to earn enough and at the same time seek further advancement. A doctor might, for instance have a private practice, a part-time hospital appointment, be a police surgeon, a poor law practitioner, and work as an assessor for a large insurance company. On this see, Anne Digby, *Making a Medical Living* (Cambridge, 1994); and M. Dupree, 'Other than Healing: Medical Practitioners and the Business of Life Assurance during the Nineteenth and Early Twentieth Centuries' (forthcoming). I am grateful to the author for an advance copy of this paper.

[5] Gibson, *RCPSG*, p. 239.

[6] See the preface to Stanley Alstead's summary of the Minutes, *RCPSG: Extracts from the Minute Books of the Faculty/College from 1798–1958 with Comments by the Compiler*, 3 vols, RCPSG, 1/13/3/1.

[7] Ibid., passim.

potential Medical Officers of Health (MOsH), was secured when the three Scottish medical corporations united to grant the qualification as a conjoint board from 1 August 1892.[8] Most medical authorities had instituted similar qualifications in the late 1870s or early 1880s, mindful of the potential market from the public service which was expanding as the state took on increased responsibilities in public health. Edinburgh and Glasgow had obtained authority to appoint MOsH early and the first incumbents took up their duties in 1862 and 1863 respectively.[9] In 1875 the Public Health Act made it compulsory for English local authorities to appoint MOsH, and in 1886 the GMC recognised the DPH. In 1888 the Local Government Act made it compulsory for every MOH to hold a DPH, legislation that was extended to Scotland the following year.[10] The Edinburgh Physicians had instituted a diploma in 1875, and the University of Glasgow in 1876. The Faculty followed suit in 1883. The Faculty's qualification was modelled on that of the RCPEd. Candidates were examined in Chemistry, Physics and Meteorology, Sanitary Medicine, Practical Sanitation, Sanitary Law and Vital Statistics. The list indicates how single-mindedly the qualification was geared towards MOH or other public service.[11] In early 1891 a novel arrangement was entered into with the University and J. B. Russell, the MOH for Glasgow. The University's courses in public health were to be opened to those taking the Faculty's diploma, and Russell agreed to take two pupil assistants for six to twelve months so that they could learn under him in an apprenticeship system.[12] In July 1892 the DPH curriculum was altered to include more practical laboratory work.[13]

These were momentous years in Glasgow for the growth of scientific medical disciplines which had a service role within the medical and the wider community. In 1895 the University opened a Pathological Institute at the Western Infirmary and appointed Professor Joseph Coats, a Fellow of Faculty and the main driving force behind the project, to be its Director.[14] The

[8] *Faculty Minutes*, 3 November 1891.

[9] The first Scottish MOH was Edinburgh's Henry Duncan Littlejohn (1862). Glasgow City Council had obtained authority to appoint a MOH under the Glasgow Police Act of 1862, and William Tennant Gairdner took the post on a part-time basis in 1863. James B. Russell succeeded him as the city's first full-time MOH in 1871. See David Hamilton, *The Healers*, pp. 200–7.

[10] Anne Crowther and Brenda White, *On Soul and Conscience: The Medical Expert and Crime: 150 Years of Forensic Medicine in Glasgow* (Aberdeen, 1988), p. 28.

[11] See Coutts, *History of the University of Glasgow*, p. 576; Craig, *History RCPEd*, pp. 529–33; and *Faculty Minutes*, 3 September 1883, RCPSG, 1/1/1/10.

[12] *Faculty Minutes*, 5 January and March 1891, RCPSG, 1/1/1/11.

[13] Ibid., July 1892.

[14] See, 'Inauguration of a New Pathological Institute in Connection with the Western Infirmary and the University', *GMJ*, 46 (1896), pp. 357–68.

Glasgow Corporation opened a Bacteriological Laboratory as part of the Sanitary Chambers in 1899. The municipality had accepted wider responsibilities for the isolation and free treatment of all infectious diseases occurring within the city in April 1881. To fulfil this remit a central laboratory to which the city's general practitioners could send samples for testing was needed, and it was a Fellow of Faculty, A. K. Chalmers, then a junior MOH and later to succeed Russell as full MOH, who had initiated the campaign to build a bacteriological laboratory in 1895.[15] The University had also built new laboratories for public health teaching. These had been pressed for by the new Professor of Forensic Medicine and Public Health at the University, and an active Fellow of Faculty, John Glaister.[16] Glaister and Russell took pains to make sure it was a modern facility. During 1898 they visited laboratories in London, Cambridge, Newcastle and Edinburgh.[17] When the new building was opened in 1907 it included a large toxicological laboratory and a public health section with a chemical and a bacteriological laboratory, and was the best equipped in the whole of Britain.[18] This University laboratory building programme may have been one factor in the decline of the extra-mural medical schools. They could no longer compete for students with the better equipped universities.

The Scottish medical corporations were also forced to consider closer cooperation in 1887 when the information that the English Colleges were seeking powers to confer degrees in Medicine and Surgery caused, 'a sudden flutter in the[ir] dovecotes'.[19] As part of the long process of reform of the disparate nature of higher education in London, various plans were put forward between 1884-87 for the creation of a unified London teaching university which would be able to award medical degrees. One scheme came from the London Royal Colleges and proposed the creation of a conjoint Senate of Physicians and Surgeons with degree awarding powers. A petition went forward to the Privy Council in 1887 seeking authority for this plan. The idea seems to have arisen partly because the London Colleges were worried that their qualifications of MRCS and LRCP were considered of less value than Scottish university medical degrees.[20] News of this sent the Scottish

[15] Robert M. Buchanan (City Bacteriologist), 'Bacteriology' in *Municipal Glasgow: Its Evolution and Enterprises* (Glasgow, 1914), pp. 247-53; *British Medical Association: The 90th Annual Meeting, Glasgow, July 1922. The Book of Glasgow* (Glasgow, 1922), pp. 142-44.

[16] John Glaister (1856-1932). He had previously (from 1881) been Lecturer in Medical Jurisprudence and Public Health at the Royal Infirmary Medical School/St Mungo's.

[17] John Glaister, 'Forensic Medicine and Public Health', *The University of Glasgow: Its Position and Wants by Members of the University* (Glasgow, 1900), p. 34.

[18] Crowther and White, *On Soul and Conscience*, p. 33.

[19] Craig, *History RCPEd*, p. 601.

[20] See Cooke, *History RCPL*, pp. 931-52; on the latter point see p. 937.

corporations into something of a panic and joint consultations were hastily arranged. A memorandum sent to Fellows of all three corporations recommended sending to the Privy Council a speedily drawn up counter-petition and charter, modelled on the London petition and seeking the same degree awarding powers for a Scottish 'Senate of Physicians and Surgeons'. This was to be composed of the President and six other Fellows of each corporation, and representatives of Glasgow and Edinburgh extra-mural teachers of Medicine. Additional members would be elected by the graduates when two hundred had passed. The memorandum explained the seriousness of the threat from the aspirations of the London Colleges:

> We scarcely need to point out to you that a movement of this kind, if successful, will be fraught with momentous consequences to the three Scottish Medical Corporations which at present cooperate to form a qualifying board under the Medical Acts. It is evident that if the Scottish Bodies are limited to the exercise of their existing powers, while the English Colleges succeed in obtaining such an important addition to their present chartered rights, the balance between the Medical Corporations of these two divisions of the kingdom will be disturbed. The advantages which the English Colleges would reap would be great and cumulative. While still retaining their present distinctive character and privileges as Licensing Authorities, their status as such would be raised absolutely, and also relatively to the other Bodies, and they would have, in addition, what would be to all intents and purposes University rank. To students the attractive power of their present conjoint examination would become irresistible; for it would not only, with a good title, admit to the Register, but open a door beyond for obtaining the coveted degree. What would then be the effect on the examinations of the Triple Board and the colleges cooperating to form it? The draining away of the present supply of candidates to an extent to which it would be difficult to fix a limit would greatly diminish their importance as Licensing Bodies, while their status would ... also suffer relatively from that of the English Colleges being raised.[21]

The Scottish Colleges dared not oppose the London plan: for if it was approved how could they then have claimed the same powers for themselves?[22] Thus the counter-petition in favour of a degree-awarding Scottish senate of the corporations was adopted and sent to the Privy Council in January 1888.[23] All the other British universities, the colleges that would

[21] 'Memorandum to the Fellows of the Royal College of Physicians of Edinburgh, of the Royal College of Surgeons of Edinburgh and the Faculty of Physicians and Surgeons of Glasgow', from R. Peel Ritchie, Chairman of Meeting of Delegates, 14 December 1887, p. 1, RCPL, Muniments, Misc. Papers, 380.

[22] Ibid., p. 2.

[23] Craig, History RCPEd, p. 602.

become the new civic universities and the corporations opposed the plans of the London Colleges,[24] and the Scottish petition was dropped, having done its defensive work.

As well as showing how the British medical corporations operated in a state of constant jealous tension, rivalry and competitiveness in this period, the episode does illuminate the ongoing competition for control of medical education in Glasgow between the University, and the Faculty and the extra-mural schools. The University degree, which the London Colleges sought to replicate, was fast becoming the first qualification of choice for intending medical practitioners. This was a battle that the University was to win after 1948, when it assumed control of all undergraduate education, but, before that date, it is important to remember that there was a battle for students in Glasgow between the Anderson's and the Royal Infirmary/St Mungo's Medical Schools, with their strong affiliation to the GRI and students who largely took the TQ, and the University with its GWI. There was also, however, much interchange of students between the two, at least early on, when a medical education could still be assembled piecemeal from a number of different institutions. This is an area in which further work is needed. Did the University and the extra-mural medical schools evolve distinctive styles of medical education, the former perhaps more academic and theoretical, and the latter more practical and clinical? Certainly the teaching staff of the extra-mural schools were mostly Fellows of the Faculty. Their students took the TQ, and, like the Faculty itself, the extra-mural lecturers represented the local clinical elite. Or did the extra-mural schools deliberately seek to provide a particularly progressive and scientific medical education as a selling point in the competition for students with a more traditionally-minded University, at least until they could no longer compete in the provision of expensive laboratory facilities?

In any case, the Faculty's response to the attempt by the London Colleges to obtain degree awarding powers demonstrates the narrow, short-term thinking of its officers in this period. Once the initial scheme had been thwarted, nothing was done to address the underlying problem of the rise of the university medical degree, which was eventually to break the axis of the Faculty, the GRI and the extra-mural schools and leave the Faculty facing the prospect of having only a very limited professional role and income.

There was a renewal of an ancient activity in the 1880s and 1890s when

---

[24] The exact positions of the various medical authorities on the London scheme are given in, 'Senate of Physicians and Surgeons Charter. Copy Letter: The Clerk of the [Privy] Council to Mr Lock', 17 February 1888, Glasgow University Archives and Business Records Centre (GUABRC), 58591.

the Faculty Minutes record numerous Faculty investigations into alleged unethical practice by Glasgow doctors. The growth of a mass market in Britain had led to the commercialisation of increasing numbers of quack remedies. Manufacturers of these pills and potions avidly sought the endorsement of registered medical practitioners to add weight to their extravagant claims as panaceas. The Faculty became directly involved in cases involving its Fellows, often on its own initiative rather than that of the GMC, and thus had a role in the evolution of professional medical ethics. In 1886 one Dr Robert Bell, a Fellow, had become professionally involved with the production of a pharmaceutical preparation, to the extent that it was named 'Dr Bell's Liniment'. Described in *Faculty Minutes* as a 'secret remedy' it had clearly been advertised under Bell's professional authority. Faculty considered his conduct 'infamous in a professional respect' and suspended his Fellowship for six months. This case led to a change in Faculty regulations in April 1887, when the reform-minded John Glaister had a motion unanimously passed that Fellows or Licentiates who were in any way associated with the commercial production and sale of quack remedies (which basically meant remedies produced commercially rather than those administered in hospital or in general practice by registered practitioners), would be deprived of his qualification and struck off the Faculty's lists.25 This was a period when the medical profession was only just coming to terms with the commercial production of therapeutic agents, and the professional approval of such agents had not yet been codified into the modern system of clinical trials for commercial products.

An even more interesting case was considered by the Faculty in 1897, that of Mr Edward Coyle. Coyle had published an anonymous pamphlet entitled 'A New Theory of Cancer' which argued that the *sole* cure for the disease was the injection of pure serum from the blood of a living animal directly into the patient's tissues. The late nineteenth and early twentieth centuries were the high-water mark of serum and vaccine therapy.26 The developing medical sciences of bacteriology and immunology seemed to promise a new therapeutic armamentarium against many infectious diseases. Building on earlier work by Pasteur, Elie Metchnikoff, Denis and Leclef, Paul Ehrlich

---

25 *Faculty Minutes*, 4 April 1887, RCPSG, 1/1/1/11.
26 This section is based on W. F. Bynum, *Science and the Practice of Medicine in the Nineteenth Century* (Cambridge, 1994), pp. 158–68; R. H Shryock, *The Development of Modern Medicine* (London, 1948), pp. 240–48; Paul Weindling, 'From Medical Research to Clinical Practice: Serum Therapy for Diphtheria in the 1890s', pp. 72–83, in John Pickstone (ed.), *Medical Innovations in Historical Perspective* (Houndmills, 1992); and Michael Worboys, 'Vaccine Therapy and Laboratory Medicine in Edwardian Britain', ibid., pp. 84–103.

and many other investigators were discovering that blood contained various elements some of which destroyed disease cells, particularly when the host had been immunised. There was a debate about whether this process was accomplished at a cellular level or chemically in the blood serum. Denis and Leclef discovered in 1895 that both elements were involved: chemical components of the blood worked on bacterial intruders making them more susceptible to white blood cell phagocytes. This process could be stimulated by immunisation with antitoxin, immunised animal blood containing specific induced antibodies. Ehrlich produced a rational theory of how antitoxin worked and from this developed an explanation for immunity to infection as a chemical combination of antigens and antibodies, for which he received the Nobel Prize in 1908. Meanwhile, Behring injected blood serum from animals that had been immunised with the diphtheria toxin (thus producing an antitoxin) into Berlin babies and seemed to have found a cure for diphtheria. Such successes led to the mass production of vaccines and serums for a variety of diseases in the Pasteur and Koch Institutes. There was little regulation of the use of such treatments, and although they were not widespread in British general practice by the early 1900s, they were quite widely advertised in the medical press.

Such developments had created a widespread public interest in the promise of this new scientific medicine, even before the rise to prominence of Sir Almroth Wright, who championed the ability of vaccines to cure as well as protect against disease in the early 1900s. Wright was lampooned by Bernard Shaw in *The Doctor's Dilemma* as Sir Colenso Ridgeon with his cry of 'Stimulate the Phagocytes!'. At a more prosaic level, Edward Coyle was exploiting the public hopes that the new methods might even find a cure for cancer, whose aetiology remains far from clear and may still include infective influences in some types. Coyle had undertaken to treat a patient in this manner after having given her (or her husband) a copy of his monograph. He asked for £20 for the service of which he took £10 in advance. On the patient's death, the husband sued Coyle for not delivering the promised cure and Coyle raised a counter action for the remaining part of his fee. The action was settled out of court with Coyle agreeing to return £5 of the pre-payment. Unlike Bell, Coyle avoided any more serious punishment for this 'improper conduct' than censure by the Faculty, but then he had apologised to the Faculty immediately and Bell had had his apology wrung out of him after numerous interviews.

While Coyle was advocating serum therapy, the Faculty had ceased to be a vaccination station in the second half of 1896, as the numbers of children being presented had dropped to under two hundred, which was too few to qualify it as a recognised vaccine station with either the Privy Council or the

Local Government Board, the two government bodies then charged with health matters. Thus ended ninety-five years of public service.[27]

The involvement of the GMC in these cases is not clear, although from other cases reported in the *Faculty Minutes* it appears that by the last years of the century it was more normal for the Council to inform the Faculty of wrongdoing on the part of one of its Fellows or Licentiates, rather than the other way around. The first recorded case of the GMC informing the Faculty of wrongdoing and the Faculty then taking disciplinary action appears to have been in 1894 in the case of Mr S. B. Niblett, who had been struck off the register by the Council and was similarly treated by the Faculty in September of that year.[28] During this period the GMC was still establishing its procedure in cases of medical discipline, and periodically tinkering with the bureaucratic machinery.[29] It would be interesting to know more about how remedies became recognised as orthodox as opposed to quack. The emergence of the clinical trial is central to this issue; but, clearly, social factors as well as technical ones must have come into play, since the validity of clinical trials is still often a moot professional point when doctors advise drug companies and various lobbies fund trials.[30]

The last years of the old century and the first few years of the new brought important changes for the Faculty. On 25 May 1893 the new Faculty meeting hall was opened at the very back of the 242 St Vincent Street building. The planning was done by the famous Glasgow architect John James Burnet,[31] the work being completed by the firm of which he was part, John Burnet Son and Campbell. Underneath the hall a new examination suite of four classrooms was laid out, as well as a porter's box and furnace room (this

---

[27] *Faculty Minutes*, 3 August 1896.

[28] Ibid., 3 September 1894.

[29] See Russell G. Smith, *Medical Discipline: The Professional Conduct Jurisdiction of the General Medical Council, 1858–1990* (Oxford, 1994). This account is, however, rather over-legalistic, focuses on the latter part of the period, and does not explore how drugs become accepted by the profession as suitable to prescribe or endorse.

[30] On this see, for example, Malcolm Nicolson and Cathleen McLaughlin, 'Social Consructionism and Medical Sociology: A Study of the Vascular Theory of Multiple Sclerosis', *Sociology of Health and Illness*, 10 (1988), pp. 234–61.

[31] John James Burnet (1859–1938). One of a flourishing group of nineteenth-century Glasgow architects which included David Hamilton (1768–1843); Alexander 'Greek' Thomson (1817–1875); and Charles Rennie Mackintosh (1868–1928). Burnet's other works in the city include the Barony Church in Castle Street (1886); Glasgow Stock Exchange and the Pathology Department at the GWI (1894); the Botany and Engineering Building at the University and the Alhambra Theatre (1910); and the Royal Hospital for Sick Children (1912). He had equal success south of the border, moving to London in 1905 whilst retaining his Glasgow business. He was knighted in 1914. Details form Hutchison, *Property and Environment*, p. 76.

area became the Crush Hall examination waiting-area, kitchens and cloak-rooms and has just been renovated as part of the College's 400th anniversary celebrations). The hall 'is a classical example of Victoriana with its heavy wood panels, high ceiling with curved beams and the massive fireplace at the west end'.[32] Burnet also built an imposing new staircase to lead up to the hall, topped by pillars and a decorative arch above the hall doorway. The staircase was illuminated by side windows and a roof-light which duplicated the cupola light of the original house.[33] Though the work was done very quickly it was not ready in time for the November 1892 Faculty dinner, and so, for the first and only time, the dinner was held elsewhere: next door in the Windsor Hotel. In 1899 electric light was installed for the first time throughout the Faculty building at a cost of £85 4s. 9d.,[34] though it had been installed in the Faculty Hall in 1893.

In spite of the improvements and extensions, the Faculty was still short of space, particularly for reading rooms and the ever increasing book collection. Thus in 1900 the Faculty purchased the house next door at 236/38 St Vincent Street, again under the guidance of Burnet, for £4200. The house, built around 1836, was previously in the possession of the engineer William Tait of Mirrlees and Tait, but from 1886 it was rented to Thomas P. Crawford of Crawford and Co., a steamship business. In December 1902 the Council and Finance Committee met to apportion rooms to uses in the new premises. One of the new examination rooms under the hall in 242 was used for anatomy examinations and contained dissected specimens and models, which left two rooms for other subjects of the first, second and third year examinations. This meant that two rooms had to be found in the new building for the final examination. The largest room in the new basement was fitted out as an examination room for mechanical dentistry, underlining the importance of the LDS to Faculty finances at this time.[35]

More importantly, in February 1901 James Finlayson, at that time Faculty President, announced his retirement as Honorary Librarian. He had held this post since 1877. Alexander Duncan had been Librarian since 1865, and during the years that these two served together the library was turned into a first-class scholarly resource. Finlayson undertook a survey of the holdings and identified areas of medicine where stock was thin and purchases needed to be made. Duncan attempted to fill these gaps by trawling the local

---

[32] Ibid., p. 89.

[33] Ibid., p. 86.

[34] *Faculty Minutes*, 6 November 1899.

[35] Details from Hutchison, *Property and Environment*, pp. 97–99.

second-hand shops and arranging swops with other medical libraries so as
to assemble a comprehensive collection equally representing all the departments of medicine. A great boost had been given to the library holdings with
the donation of the William Mackenzie collection in 1883.[36] This was Mackenzie's personal medical library, which, unsurprisingly, consisted mostly of
ophthalmological works. In 1885 the first catalogue of the library holdings
was finished; it had been delayed to accommodate the Mackenzie collection.
For the first time the books were arranged on the shelves by subject. The
books, periodicals, reports and transactions listed in the 1082 page catalogue
numbered about 25,000 items, compared to 8500 in 1853. By 1896 Duncan
estimated that there were around 40,000 volumes. This increase necessitated
a second volume of the catalogue which did not appear until 1901 because
Duncan was busy completing his history of the Faculty to 1850, *Memorials
of the Faculty of Physicians and Surgeons of Glasgow*.[37] This second volume
indexed acquisitions from 1885 to 1900. Duncan himself died in office in
1921.[38]

    In December 1909 the Faculty gained the title 'Royal' from the King via
the Secretary of State for Scotland. This was the work of the then President
John Glaister, the Regius Professor of Forensic Medicine at the University,
whom we observed earlier involved in the expansion of university laboratory
facilities in this period. There had been an earlier attempt, in 1903–4, to
change the name of the Faculty to the 'Royal College of Physicians and

---

[36] William Mackenzie (1791–1868). After studying arts at Glasgow University Mackenzie
switched to the study of medicine for five years and in 1813 was appointed Resident Clerk to
Dr Richard Millar at the GRI. He became a Licentiate of the Faculty in 1815. He went to
London and studied at the Eye Infirmary and at St Bartholomew's under John Abernethy,
before embarking on a grand medical tour of Europe during 1816–18, where, among others,
he observed the work of Dupuytren, and bought much of the library that would later be
presented to the Faculty. Returning to London, he obtained the Membership of the RCSEng,
and became a general practitioner, while also running lecture courses on eye diseases (he had
started to specialise in this area while a student). He returned to Glasgow as Professor of
Anatomy and Surgery at Anderson's College Medical School in 1819 and was instrumental in
the founding of the Glasgow Eye Infirmary in 1824. In 1830 he published his first major work,
A Treatise on Diseases of the Eye. This was followed in 1841 by *The Physiology of Vision*. In
1828 he had been appointed Waltonian Lecturer in Diseases of the Eye at Glasgow University,
graduating MD in 1833. He became Surgeon Oculist to the Queen in Scotland in 1838. He
was also a very active member of the Faculty (as Examiner, Sealkeeper, GRI Manager, and
Visitor in 1850). He was elected President in 1855, but declined the office on health grounds.
See James Beaton, 'William Mackenzie: A Short Biographical Essay', unpublished paper, 1997.
I am grateful to the author for a copy of this paper.
[37] Alexander Duncan, *Memorials of the Faculty of Physicians and Surgeons of Glasgow* (Glasgow, 1896).
[38] Details here from Gibson, *RCPSG*, pp. 202–9.

Surgeons of Glasgow', instigated by Ebenezer Duncan,[39] but this had been voted down after strong opposition from two Faculty heavyweights, Hector Clare Cameron and James Finlayson. Alstead commented on this that:

> Had the project succeeded, there is little doubt that the change of name would have had an important effect on the status and health of the Corporation during the following half century a period marked by a diminution in its influence in the medical life of Glasgow and Scotland.[40]

The Faculty initially believed that an Act of Parliament would be needed to update the name on qualifications already registered with the GMC, but in the event the GMC made no trouble and simply accepted the change for all official purposes in early November 1910. The achievement of the new title was celebrated in a grand *conversazione* on 28 October 1910 at which, before 620 invited notables, a striking mace was presented to the Faculty by Dr James Walker Downie. Downie had been a Fellow since 1884, was a general practitioner and Dispensary Surgeon for Diseases of the Throat and Nose at the GWI, and was also attached to the Royal Hospital for Sick Children. He later held a University lectureship and served for a year as Faculty Visitor in 1915.[41] The mace, which, carried by the College Officer, still precedes the President and Office Bearers into Meetings of the College, and is laid on the table when the College is in session, was modelled on that of the House of Commons, and was made by Frank Lutiger of West Kensington. It has four faces depicting the royal emblems of James VI and Edward VII and the arms of Glasgow and of the Faculty. The new Faculty arms adopted

[39] Ebenezer Erskine Duncan (1846–1922). Duncan was born in Renton, Dunbartonshire, and graduated MB, CM from Glasgow University in 1867. He obtained a post as house surgeon to Lister in 1867, the year that Lister was publishing his first papers on the antiseptic principle. Duncan obtained his MD in 1870, and in 1871 was elected a Fellow of Faculty. He set up a general practice located, by 1869, in Crosshill, Glasgow. He joined the Southern Medical Society (becoming its President in 1877). From 1878 he began to campaign for a hospital to serve the south side of the city. He was thus one of the founding fathers of the Victoria Infirmary which opened in 1890, and he became a visiting physician to the new hospital. Duncan published *inter alia* epidemiological studies of local typhoid fever outbreaks, and on tuberculosis, alcoholism and food adulteration. He became Medical Officer of Health for Crosshill, was involved in medical police and insurance work and was often a referee in medico-legal disputes. He became Lecturer and then Professor of Medical Jurisprudence at Anderson's College Medical School, and it was on his motion that the Faculty instituted a Diploma of Public Health in 1883. He was President of the Faculty from 1915–18. See Stefan Slater and Derek Dow (eds), *The Victoria Infirmary of Glasgow, 1890–1990* (Glasgow, 1990), pp. 134–39.
[40] Alstead, *Extracts*, December 1903 to Spring 1904.
[41] MacQueen and Kerr, *The Western Infirmary, 1874–1974* (Glasgow, 1974), p. 9; Hutchison, *Property and Environment*, pp. 146–47.

in 1863 were also officially 'matriculated' or registered with the Lord Lyon King of Arms to further mark the occasion. In 1965 at the funeral of Sir Winston Churchill, when the College was represented by the then President, A. B. Kerr, the mace was borne in St Paul's Cathedral by the College Officer, Mr E. Brown.[42]

Not all Glasgow medicos were impressed with the new title however. Alstead reports that Ralph Stockman, the Professor of Materia Medica at the University, scoffed at the honour, adding, 'but, of course, I am a Republican'.[43] Stockman had however deigned to become a Fellow in 1898 by the *ad eundem* method, as an existing FRCPEd.[44] This was just after he had become a professor with wards in the GWI, in 1897. There was some controversy about his getting of these wards, and perhaps he had applied for the Fellowship to smooth things. All the other visiting physicians at the hospital in this period were Western Infirmary men, having risen through the ranks to become chiefs. By taking the Glasgow Fellowship, Stockman was attempting to establish links with the Glasgow clinical elite. This episode illustrates that although the Fellowship was not highly thought of it was still helpful to have it if wishing to gain local hospital appointments.[45]

There were also disturbing precedents which dimmed the celebratory atmosphere. In 1901 the Carnegie Trust had announced its intention to allocate £2 million to fund students attending Scottish universities. The President, James Finlayson, wrote in earnest to the trust about 'certain considerations which may have a bearing on the form or range of [the] benefactions', namely the existence of the extra-mural colleges and the Faculty's qualifications as alternative undergraduate examinations. Finlayson tactfully pointed out that grants to university students only might 'destroy' the delicate balance between the universities and the extra-mural schools in medical education, or, rather, that:

> The Extra-Mural Medical Schools could not sustain their competition with the University Schools were the students of the latter placed at an advantage over those of the former in the matter of exemption from payment of class fees; and the Licensing Bodies, such as this Faculty, would equally suffer from the draining away of the students of the Extra-Mural Schools ...[46]

The letter asked that both sets of students be treated as part of the one university system, and took the line that the money the Glasgow and Edinburgh

[42] Hutchison, *Property and Environment*, p. 150.

[43] Alstead, *Extracts*, p. 508.

[44] *Faculty Minutes*, 5 September 1898, RCPSG, 1/1/1/12.

[45] See MacQueen and Kerr, *The Western Infirmary*, pp. 45–46, 75–76.

[46] James Finlayson, PFPSG, to Andrew Carnegie, 4 June 1901, *Faculty Minutes*, 1 July 1901.

Corporations earned from such students taking the TQ went to support the upkeep of museums and libraries for students. By July more details of the Carnegie scheme had come out and it appeared that while university students' class fees were to be paid, their graduation fees would not be covered, which was of some help in somewhat limiting the now otherwise irresistible attraction of the university degree in Medicine.[47] The extra-mural schools' museums and libraries might also attract separate grants as educational resources, but the students would not get grants for attending these schools.

The rise of university undergraduate qualifications in medicine and the corresponding decline of the TQ was further exacerbated in the early years of the twentieth centuries by the growth of the English 'civic universities' such as Birmingham, Bristol, Leeds, Liverpool, Manchester, Sheffield and Wales, all of which developed medical schools offering new degrees in direct competition with the qualifications of the Scottish corporations.[48] As we noted above, the income from the TQ dropped substantially after 1900 to an annual figure of around £400.

A dwindling role in undergraduate education might have resulted in a reexamination of the function of the Faculty in this period. Did the Faculty, for instance, become involved in the drive towards social reform which characterised this period of increased municipal activity to improve the public health in the context of a national climate of concern with 'National Efficiency' most clearly marked by the Report of the Interdepartmental Committee on Physical Deterioration of 1904?[49] In December 1903 the Faculty received an invitation from the RCPEd to join its investigation into the health of urban school children by appointing a Glasgow-based committee of enquiry, with the intention of pooling the findings in a joint report.[50] The Edinburgh College had been prompted to begin its own investigations after the publication of the Report of the Royal Commission on Physical Training in Scotland in 1903.[51] The Faculty agreed and appointed a committee consisting of the President, Henry E. Clark, the Visitor, Dr W. L. Reid, and Finlayson, Glaister, Henry Yellowlees, A. Sloan and A. K. Chalmers, the latter a tuberculosis specialist who had succeeded Russell as Glasgow MOH in 1898. The Edinburgh Surgeons also became involved, and the joint report of early 1904 clearly demonstrated its provenance in its

[47] Ibid.

[48] See David R. Jones, *The Origins of the Civic Universities* (London, 1988).

[49] On this see G. Searle, *National Efficiency: A Study in British Politics and Political Thought, 1899–1914* (Oxford, 1971).

[50] *Faculty Minutes*, 7 December 1903.

[51] Craig, *History RCPEd*, pp. 732–34.

opening statement that accurate data would be of great use in 'determining the question which is at present disturbing the public mind viz.: Whether degeneracy of the race is taking place'. The report suggested a scheme for the examination of schoolchildren at progressive ages which included observation of height, weight, chest measurements, muscular strength nutrition and general development, diseases and malformations, malfunctions of organs of special senses, mental development, and the condition of the teeth. All tests were to be carried out to the highest level of scientific accuracy, and the medical corporations agreed to partially fund the study, although they would not take responsibility for its completion themselves[52] However the joint committee had got no further than this statement of intent to support an investigation by early 1905, at which time plans were in any case postponed in order to see if the forthcoming Scottish Education Bill addressed these issues. In the event the Bill was very satisfactory containing provisions for the medical inspection of schoolchildren and for the introduction of compulsory physical training. Local authorities were permitted to provide free school meals in 1906, and the passing of the Scottish Education Act in that year launched the School Medical Service.[53]

It is difficult to tell how much the joint committee report influenced matters. Probably of greater importance were the efforts of Chalmers outside Faculty. He initiated a contemporaneous enquiry by medical specialists into the health of Glasgow schoolchildren, perhaps frustrated by the slow pace of Faculty activity. This report which was sent to the City Council disclosed unexpectedly high levels of neglected eye and ear defects; dental decay in 87 per cent of the sample; past rickets in a fifth of the poorest children, many of whom were badly deformed, undernourishment in 16 per cent of the sample; and widespread infestation with lice or fleas.[54]

However fleeting or superficial this interest in child health was, it was at least some involvement. The early years of the twentieth century saw few other such forays into social policy questions. In February 1904, on the invitation of the City Council, the Faculty resolved to support its effort, in combination also with the Chamber of Commerce, to bring pressure on central government to advance preventative measures against malaria on the west coast of Africa and the tropical colonies of Great Britain. Alstead commented wryly that, 'The ravages of poverty, malnutrition, appalling

[52] Details and quotation from the report are from ibid.
[53] Ibid.
[54] See Sir Alexander MacGregor, *Public Health in Glasgow, 1905–1946* (Edinburgh, 1967), p. 81. See also W. Hamish Fraser and Irene Mavor, 'The Social Problems of the City', pp. 352–93, in Fraser and Mavor (eds), *Glasgow*, ii, *1830–1912* (Manchester, 1996).

slums and infectious diseases in Glasgow seem, in retrospect, to have been much more relevant to the Faculty's mission in the city'.[55]

Indicative of the lack of interest of the Faculty in social issues in this period was the fate of a proposal to set up an enquiry into the food habits of the labouring classes in Glasgow. This was put forward as a motion by Chalmers, who was obviously doing his best to direct the Faculty towards a more active role in social policy, in December 1906. Seconded by Dr J. Wallace Anderson, Chalmers wanted the Council to decide the form of the survey and the district of the city to be studied. However the proposal was scotched by eleven votes to five at the Faculty meeting on 7 January 1907, after opposition from W. K. Hunter.[56] The study did go ahead, however, under other auspices and the results were published by D. W. Lindsay in 1913 as a *Report upon a Study of Diet of the Labouring Classes in the City of Glasgow*. Also voted down at this meeting was a motion to request the GRI Governors to include wards and dispensary facilities for the treatment of mental and nervous diseases as part of their rebuilding.

In February 1907 Chalmers again tried to oppose reactionary elements within the Faculty. He argued strongly against a motion stating that a new public dispensary for the poor (the Anderston's Health Association) was unnecessary, as there was already enough medical support for local inhabitants. Chalmers urged his colleagues to consider the position of the poorer classes, but his motion was resoundingly defeated by twenty-four votes to his one. At the same meeting a Faculty enquiry into the perceived abuse of charitable medical assistance in Glasgow by the poor was established instead.[57]

This lack of interest in social problems may indicate that the Faculty believed that other bodies were already working in this area, notably the City Council. Chalmers was attempting to mobilise the Faculty into supporting these campaigns as an authoritative figurehead institution representative of elite Glasgow medical opinion. Faculty reluctance may simply have been the understandable inertia of a body that saw itself purely as a professional *licensing* body, whose origins lay in the guild system and whose aim was the protection and maintenance of skills, to become involved in controversial social action. There may also have been a political element, a reluctance to associate the ancient (and thus naturally conservative) professional body with so-called gas and water socialism. As Callum Brown has argued, in the early years of the twentieth century, 'the Faculty refused to become a pressure

55 Alstead, *Extracts* (commentary on *Faculty Minutes* February 1904), p. 475.
56 *Faculty Minutes*, 3 December, 7 January 1907.
57 *Faculty Minutes*, 4 February 1907, RCPSG, 1/1/1/13.

group, preferring to maintain autonomy and impartiality as a professional body'.[58]

The one issue which did activate the Faculty in the early years of the twentieth century was that of National Health Insurance. The Faculty approached this, as it was to approach the National Health Service which many commentators have seen as foreshadowed in the earlier welfare reforms, as a question of the infringement of the rights, privileges, status and income of the medical practitioner.

The introduction of the National Health Insurance scheme, partly modelled on Bismarckian social welfare schemes in Germany, by Lloyd George as Chancellor of the Exchequer in 1911 was a period of great concern for medical practitioners.[59] General practitioners were worried that they were going to be dragooned into a state medical service as salaried civil servants under the control of local authorities, thus losing their coveted professional independence. Consultants were concerned that the reform of general practice might lead to greater privileges for GPs (for instance specialisation and hospital work) which might threaten the socio-economic status of the consultant. This professional jealousy between consultants and GPs was of long-standing and the division of the British medical profession is one of its most distinctive features and has, in part, produced the health care system we have today.

Before the Act, at least half of the GPs in Britain were engaged in contract practice with Friendly Societies. Lloyd George had to negotiate the insurance scheme with these societies, eager to retain control of their doctors, the large commercial insurance companies (called, with the Friendlies, the Approved Societies), and the medical profession. However he took greater pains with the Approved Societies than with the doctors,[60] and this may have left a legacy of distrust of the state which informed later opposition to Bevan's attempts to establish the NHS.[61] The scheme left the Approved Societies in control of deciding which doctor treated which patient; the doctors with no representation on the administrative bodies and thus no say in the scheme's conduct; and no income limit on who could join the scheme. The doctors'

[58] Callum Brown, 'The Royal College of Physicians and Surgeons and the Development of Social Policy, 1850–1920: An Introductory Discussion', RCPSG, 20/4/29, p. 17.

[59] I am grateful for the help of Malcom Nicolson in this section. Perhaps the best secondary source for the dispute over NHI is J. L. Brand *Doctors and the State* (Baltimore, 1965); but see also Frank Honigsbaum, *The Division in British Medicine* (London, 1979).

[60] On this see, John Turner, '"Experts" and Interests: David Lloyd George and the Dilemmas of the Expanding State', pp. 203–23, in Roy MacLeod (ed.), *Government and Expertise* (Cambridge, 1988).

[61] This point is made by Craig, *History RCPEd*, p. 704.

fear was that private practice would dwindle to nothing and that they would be left as the 'the hacks of the service [paid] six shillings a year for medical attention to insured persons'.[62] In England British Medical Association opposition was intense, even after the Act, urging its amendment to include the so-called 'Six Cardinal Points'. These were an income limit of £2 a week to join the scheme; patients to have free choice of panel doctors; medical and maternity benefits to be administered by insurance committees not by approved societies and questions of medical discipline to be settled by local medical committees composed entirely of doctors; method of payment by insurance committees to be decided by a majority of local doctors; adequate payment to be given (later put at 8s. 6d. a head, excluding the cost of medicines); adequate medical representation on the various administrative bodies working the Act; and statutory recognition of the local medical committees.[63]

In Scotland opposition was less intense, Hamilton thinks perhaps because of the more limited opportunities for lucrative private practice, and the publicly and professionally recognised need for such a service in unhealthy urban and rural areas,[64] but fully involved all three medical corporations. The Bill was submitted to the Faculty meeting of 5 June 1911 and at the same meeting an invitation was read from the Edinburgh Physicians asking for a Faculty representative to join a joint committee of Edinburgh Physicians and Surgeons to formulate 'a joint plan of action'. After Alexander Napier had expressed the hope that the Bill should not pass until the medical profession had been thoroughly consulted,[65] the Council was asked to consider the Faculty's position on the Bill, including the drawing up of any resolutions to Parliament for its modification.[66] The Council met on the 8 June and

[62] Paul Vaughan, *Doctors' Commons: A Short History of the British Medical Association* (London, 1959), p. 198.

[63] Ibid., pp. 201–2.

[64] David Hamilton, *The Healers: A History of Medicine in Scotland* (Edinburgh, 1981), p. 243.

[65] Alexander Napier (1851–1928). A Fellow of the Faculty in 1877, he built up a family general practice in the Queen's Park area of the city, and was elected one of the first two visiting physicians to the Victoria Infirmary when it opened in 1890 (Dr Ebenezer Duncan was the other). Napier was assistant to the Professor of Materia Medica at the University (Stockman), and then taught the subject at Anderson's College. He was an active Fellow, serving as Councillor and Member of the Library Committee, and succeeding Lindsay Steven as Honorary Librarian in 1909. He was Secretary of the Business Committee of the Glasgow and West of Scotland Medical Association in the 1880s and, with Joseph Coats, he coedited the *GMJ* as its organ from 1882–88. Though an old-school physician he was alive to modern scientific developments, being one of the first to introduce the use of tuberculin to Glasgow. See obituary, *GMJ* (1928).

[66] *Faculty Minutes*, 5 June 1911.

drew up a resolution which was amended at the Faculty meeting on the 12 June, when Dr John Glaister and Dr Alexander Napier were also appointed to act as Faculty representatives in joint meetings with the Edinburgh Colleges. The Faculty sent copies of its resolutions to Lloyd George and all Scottish MPs.[67] The resolutions echoed the BMA's six points drawn up at its special meeting in London on 1 June, which Lloyd George had famously addressed, but also made four further points: that due payments for use should be made to voluntary hospitals so that they should not suffer financial loss under the legislation; that surgical dressings as well as drugs (of British Pharmacopoeia standard) should be paid for; that medical benefit should include hospital treatment when necessary; and that medical practitioners should be paid to attend births in addition to a midwife, if she requested a medical presence.[68] Glaister and Napier attended a joint meeting of the three corporations in Edinburgh in late June when resolutions similar to those of the Faculty were agreed. On Saturday 25 November at four o'clock, just as the Bill was passing through the House of Commons and being amended there, a large meeting of Fellows, Licentiates and members of the medical profession in Glasgow and the west of Scotland was held in Faculty Hall. The meeting had been called to discuss the Faculty Council's latest resolutions that one eighth of the Scottish Insurance Committee should be medical practitioners and that voluntary hospitals should be reimbursed for any services rendered under the Bill by the Commissioners.[69] When these were put before the meeting Dr MacGregor Robertson and Dr Hamilton immediately objected that instead of such details the meeting should be discussing 'whether they approved of chaining round the neck of the profession for ever the abominable system of contract practice'.[70] Calling the meeting to order the chairman, the Faculty President, Dr James A. Adams, phlegmatically observed that:

> it was all very well to pass pious opinions about the Bill, but it would probably become law in spite of what they did. He agreed very largely with what had been said, but he desired the public to recognise that while they made this protest they were doing all in their power in a courteous and tolerant manner to consider the scheme and to do the best they could with a bad bargain.[71]

[67] Ibid., 12 June 1911.

[68] 'RFPSG: National Insurance Bill', printed copy of resolutions in *Council Minutes*, 8 June 1911, RCPSG, 1/1/3/6.

[69] Ibid., 20 November 1911.

[70] 'The Insurance Bill: Meeting of Glasgow Physicians', *Glasgow Herald*, Monday 27 November, 1911.

[71] Ibid.

The first Council resolution was then approved. Professor Glaister then spoke on the subject of the second resolution, a topic which particularly affected hospital consultants, and those GPs with hospital appointments (like Alexander Napier for instance),[72] the financial health of the voluntary hospitals:

> under this Bill each workman would have to pay so much per week and there was little likelihood ... of his contributing to any other voluntary fund. Each employer, too, would have to pay so much per week for his employees, and it was also unlikely that he would contribute. Those contributions in Glasgow formed practically two-thirds of the whole revenue, and were it not for the bequests from which deficiencies in annual expenditure were made up, the hospitals in Glasgow would be bankrupt. If the workmen's and employers' contributions were to go by the board in this fashion, there was hardly any likelihood that any money would be left for voluntary contribution, and the hospitals would accordingly lose that income which they had long enjoyed as a means of affording charitable relief to those who required it.[73]

Steven Cherry has shown that while the most important source of Scottish voluntary hospital income in this period was investment income, income from collecting and contributory schemes came a close second.[74] The Council's second resolution was then approved, and it was agreed to send the two resolutions coming out of this meeting to the GMC, the Prime Minister, the Chancellor of the Exchequer and the Glasgow MPs. The meeting then moved on to consider other aspects of the question and, after McGregor Robertson had whipped up the assembled doctors with a motion strongly condemning contract practice, a moderate motion by Dr John Duff, which praised the Bill 'as a great and beneficent measure of social reform' and recommended that doctors consent to work under the Act, was utterly defeated. The meeting ended by instead approving of a motion from Dr Pirrie and Dr J. P. Duncan that, unless the Bill was amended to include the BMA's six points, 'no medical practitioner shall submit to the rules under the Bill'.[75]

The combined National Insurance Committee of the three Scottish corporations met again in Edinburgh on 20 December, and a joint manifesto was agreed. It was published in late December and was sent to all registered medical practitioners in Scotland. This manifesto saw the Faculty and the

[72] It must be remembered that in this period when specialisation was in its infancy, there was much overlap between what later became increasingly these two distinct types of practitioner.

[73] 'The Insurance Bill', *Glasgow Herald*.

[74] Steven Cherry, 'Before the National Health Service: Financing the Voluntary Hospitals, 1900–1939', *Economic History Review*, 50 (1997), pp. 305–26.

[75] 'The Insurance Bill', *Glasgow Herald*.

other Scottish corporations for the first time moving into the sphere of public policy as professional pressure groups. All Scottish corporations were representative of consultant and general practitioner interests, as they granted both Fellowships and Memberships and conjoint and single licences. They now wished to become spokesmen for those interests. The manifesto declared that, 'Now that the Bill has become law they consider it their duty to advise the profession in Scotland as to what, under these circumstances, should be done'.[76] The Faculty was perhaps more representative of local medical opinion, and of general practitioners in particular, than the other two Scottish corporations, since it had historically included physicians and surgeons, and had long offered a complete single licence for general practice. Since 1906 the Faculty had also included on its Council a representative of the Licentiates, after pressure from the Association of Licentiates of Glasgow.[77] Yet the joint manifesto stressed that,

> The Scottish Corporations, who are cordially in sympathy with the Profession, and are anxious to assist it in any way in their power, recognise that at this critical juncture *united action is essential.* They wish it to be clearly understood that their desire is to assist and strengthen the action of the British Medical Association in Scotland.[78]

The manifesto went on to argue that since Scotland was to have a separate administrative organisation for National Health Insurance (a separate executive, commissioners and fund) the Scottish medical profession should also have 'a strong and thoroughly representative Central Medical Council or Committee, endowed with full advisory and administrative powers, with a paid secretary and an office in Scotland'. The manifesto suggested that such a medical council might consist of the Scottish Committee of the BMA; one representative from each of insurance areas to be established under the Act; and representatives from the Scottish corporations and universities, and that the corporations would contribute towards the cost of such a council. The Council was to be an independent body representing professional opinion campaigning for the profession's view of how medical work should be organised under the Act, but it was also to be a model of the kind of official body the profession thought should be included in the administrative arrangements with 'full advisory and administrative powers'. The same kinds of demands were put forward by the English BMA as part of the six points.

---

[76] 'National Insurance Act: Manifesto by the Three Scottish Medical Corporations', 22 December 1911, insert with *Council Minutes*, 26 December 1911.

[77] *Faculty Minutes*, 14 December 1906.

[78] Ibid.

In the end the profession had to settle for medical representation on the various administrative bodies working the act, and the official recognition of Local Medical Committees as part of that structure, with more than advisory, but less than executive powers. That this kind of activity was novel for doctors in general was emphasised by Dr Ewen Maclean, Chairman of the BMA's national Representative Body. Introducing Lloyd George to the Special Representative Meeting of the BMA in London on 1 June 1911, he commented that: 'We have not been called upon before to consider the pros and cons of the financial economic questions as they affect the medical profession'.[79]

In this sense there are interesting parallels with the contemporaneous 'public science' of natural scientists, expressed through bodies like the British Science Guild, that, 'body of rhetoric, argument and polemic' used by scientists to 'justify their activities to the political powers and other social institutions upon whose good will, patronage, and cooperation they depend'.[80] The British Science Guild acted as both pressure group for, and model of, the kind of national scientific advisory council that many leading scientists believed should be part of the state apparatus, giving scientific input into government policy. Like the doctors, the main reason why scientists had become interested in policy was that the state had begun to encroach on their professional domain because of the expansion of its role in the late nineteenth and early twentieth centuries. In both cases the key professions of the modern state wanted increased government funding (a state science policy meant decently remunerated scientific careers; NHI meant a realistic salary from a single source for general practitioners), but did not want increased government control. Many natural scientists wanted to control the emerging science policy. GPs feared becoming salaried civil servants, elite consultants feared a reorganised general practice encroaching on the hospitals, but also feared that NHI was the thin end of a wedge which would end in a state hospital service. Both doctors and scientists thus got involved in medical and scientific politics to ensure that the professions retained autonomy in an era of state funding. As the state's responsibilities continued to grow over time, both groups began to press for increased policy influence, partly as the ultimate mark of professional status.[81]

The manifesto ended with a radical rallying call to the Scottish profession, which echoed the tone of the end of the November meeting in Faculty Hall:

[79] Quoted in Vaughan, *Doctors' Commons*, p. 201.

[80] Turner, 'Public Science', p. 590.

[81] For more on this see my thesis, 'Passwords to Power: A Public Rationale for Expert Influence on Central Government Policy-Making. British Scientists and Economists, c. 1900–c. 1925' (unpublished Ph.D. thesis, University of Glasgow, 1994).

The Scottish Corporations are of the opinion that the Members of the medical profession in Scotland should refrain from undertaking any medical work under the Insurance Act until Regulations have been framed by the Scottish Insurance Commission which are entirely in accordance with the six fundamental requirements of the profession, and that the Scottish Insurance Commission should forthwith be informed of this. The Scottish Corporations would most earnestly and seriously impress upon the profession the paramount importance of loyal co-operation and of determined and unflinching adherence to the six cardinal points, and the necessity for thorough and minute organisation of the profession in each 'Insurance Area'.[82]

In January 1912, the Faculty Council further considered the make up of the 'General Council' which was to become the Scottish Medical Insurance Council, the professional body, mooted above, that was to publicise the doctors' view of arrangements, and was an autonomous replica of the similar London-based body. The Faculty Council now suggested that it should be composed of the Councils of the three corporations, representatives of the four universities, the Scottish Committee of the BMA, plus coopted representatives from urban and rural areas chosen by the medical profession in those districts.[83] The Faculty was to give £300 over two years to help form and run the Scottish body. After some heated debate in Faculty as to whether the General Council as presently constituted would be representative enough of general practitioners to secure their confidence in it, the Faculty agreed to send representatives, and of the five that the Faculty was entitled to nominate the Council elected three (Dr James Alexander Adams, the President, Dr John Barlow, the Visitor, and Alexander Napier) and the Faculty two (Mr J. P. Duncan and Mr J. MacGregor Robertson).[84] On 10 December 1912 a meeting of Glasgow doctors voted not to work under the Act as the corporations' manifesto had advised.

The Scottish Medical Insurance Council proved to be short-lived however (the Scottish BMA severed connections in May 1913),[85] as did the general influence of the corporations in health insurance. In a nation-wide BMA vote on 14 December 1912, eight out of sixteen of the Scottish divisions supported the scheme (compared with only two out of eighty-eight in England and Wales). In Scotland voting was consistent with the amount of industrial, urban practice carried out, so that large majorities in favour were recorded in

[82] 'Manifesto by the Three Scottish Medical Corporations'.

[83] Council Minutes, 11 January 1912. The English body was the Insurance Acts Committee, formed from the BMA and the Local Medical Committees set up under the Act.

[84] See Faculty Minutes, 15 January and 5 February 1912; and Council Minutes, 30 January 1912.

[85] Faculty Minutes, 2 June 1913.

Ayrshire and in Dundee, while in Glasgow only a small majority voted against. In Edinburgh, Perth and the Borders the scheme was strongly opposed.

The Scottish corporations attempted to keep the opposition alive by issuing a last-ditch manifesto in the form of a circular letter which was sent to all members of the profession in Scotland, after being approved by all three bodies. After a 'mixed reception' the manifesto was finally approved in an amended form by the Edinburgh Physicians (the Surgeons had already approved it), and was considered at a special meeting of Faculty on 16 December 1912. In the light of the acceptance of the scheme by most Scottish GPs, the letter reads like an apologia for the actions of the corporations, attempting to stress that they were concerned solely with the interests of general practitioners, and not the interests of the consultant elite. The letter began by emphasising that the reason the corporations had become involved in the Scottish Medical Insurance Council was 'an earnest desire to help the profession to maintain its independence'.[86] They had wanted to help obtain modifications of the Act in line with the BMA's cardinal points, 'to maintain and extend the freedom of the profession from lay control in the conduct of its professional duties'.[87] This was the reason for seeking adequate medical representation on the administrative bodies and for demanding statutory recognition of the local insurance committees, with more than an advisory status. The letter noted, however, that these committees were still largely made up of insured persons and not the profession, and that apart from their representation on these committees, 'the profession has no *statutory power*'.[88] The letter ended by accepting that most doctors would join up, but promising continued support for their grievances:

If members of the profession in Scotland are satisfied with such minor concessions as have been promised, and if they are prepared to accept service under Local Insurance Committees on the terms indicated, the Royal Colleges and the Royal Faculty, while not attempting to dissuade them, cannot be parties to an agreement which is derogatory to the status of the profession. In the view of the Colleges and the Faculty, the Act will perpetuate and extend some of the worst features of club-practice. The Royal Colleges and the Royal Faculty are assured that many practitioners throughout Scotland believe with them that the Act is still derogatory to the profession, and they wish it to be understood that the profession in adhering to that view may count on their sympathy and support.[89]

[86] Ibid., 16 December 1912, 'National Insurance Act: *Urgent* and *Important* Communication from the Royal College of Physicians and the Royal College of Surgeons of Edinburgh Respecting the Above Act', 'Circular Letter' to the Scottish profession.

[87] Ibid.

[88] Ibid. Original emphasis.

[89] Ibid.

After a debate in which some Fellows dissented, the President (himself a Physician at the GRI who had admitted to his own 'strong views' on the question) was empowered to sign up to the letter in the name of the Faculty.[90]

It seemed, however, that the corporations and their consultant elite were more concerned with policy influence (and perhaps also protecting their position in the hospitals from stronger, reorganised GPs) than were the rank and file practitioners. On 26 December the Scottish Insurance Commissioners announced that seventy Glasgow doctors had accepted their invitation to join the scheme. This finally broke the already crumbling medical resistance and a further meeting of Glasgow doctors was equally spilt for and against. The chairman used his casting vote in favour of the scheme and Glasgow doctors rushed to sign up, as did the doctors of most Scottish towns, only Edinburgh and Aberdeen holding out till 1913.[91]

The involvement in policy issues that the Faculty had thought necessary over NHI, and the growing realisation that the increased state involvement in social questions required a corresponding medical involvement to guard against lay control of health issues, meant that the Faculty began to develop a closer interest in such matters after 1911. This increased involvement in health politics became a growing trend as national proposals for health care policy were debated throughout the 1920s and 1930s, culminating in the close involvement in official discussions for a national health service in the 1940s. The hospital and consultant based form of the NHS meant that the Faculty developed a new role in postgraduate education which ensured an ongoing involvement in national discussions on medical education and hospital staffing.

Between 1911 and 1913 the Faculty became involved, for instance, in pressing the City Council to adopt stronger anti-tuberculosis measures.[92] In 1911 a Faculty committee was appointed to explore the issue of encouraging the Glasgow Corporation to clear more slum areas, 'the breeding places of consumption', and to provide hospital accommodation for advanced cases to isolate them and thus halt the spread of the disease.[93] The committee's 23 September 1912 report was discussed at the Faculty meeting on 7 October. The report praised the corporation for the slum clearance it had already achieved and pressed it to continue this action and to provide hospital space

[90] *Faculty Minutes*, 16 December 1912.

[91] The preceding paragraph is largely based on the account of these events given in Hamilton, *The Healers*, p. 244.

[92] For the background to tuberculosis in Glasgow, and in Scotland generally, see Neil Munro McFarlane, 'Tuberculosis in Scotland' (unpublished Ph.D. thesis, University of Glasgow, 1990).

[93] *Faculty Minutes*, 3 April and 1 May 1911.

for advanced cases. The Faculty was to send a deputation to the City Council to urge these views on them.[94] On 6 January the Faculty also decided to press for the corporation to provide alternative housing, 'structurally adapted to the requirements of healthy-living', for those families displaced by slum clearances.[95] On 20 January the Faculty added a further point for its deputation to urge on the City Council: that the owners of tenements should appoint public health instructors to advise the poor inhabitants on clean-living.[96] The Faculty deputation, consisting of President, Visitor, and the original members of the committee (Drs Knox, Chalmers, Napier, Ebenezer Duncan and W. L. Reid), was received by the Glasgow Corporation on the 6 February and found its views favourably received. The Faculty were hopeful of good results and resolved to watch developments closely.[97] However, the Faculty did later refuse to contribute funds to the City Council's campaign to establish a TB research institute.[98]

The Faculty also became more involved in giving evidence to government enquiries. In April 1914 it gave evidence to the Royal Commission on Venereal Diseases and during the First World War it became involved in debates about the mobilisation of doctors.[99] During the influenza epidemic at the end of the war, the Faculty, again at the instigation of Chalmers, sent a communication to the Food Controller urging that flu patients be allotted extra allowances of protein rich foods.[100] The war's end also brought a temporary new source of revenue as the Faculty Hall was rented out for Medical Appeals Tribunals.[101] In January 1919 the Faculty nominated members to serve on the Medical Advisory Committee on Silicosis.[102] The Faculty was also active during the 1920 outbreak of smallpox in Glasgow. William Snodgrass, then an assistant to Dr R. Barclay Ness and extra dispensary physician at the GWI, with a new general practice in the west end of the City,[103] alerted the Faculty

---

[94] Ibid., 7 October 1912.

[95] Ibid., 6 January 1913.

[96] Ibid., 20 January 1913.

[97] Ibid., 3 March 1913.

[98] Callum Brown, 'RCPSG, and the Development of Social Policy', p. 16.

[99] *Faculty Minutes*, 5 January 1914, 6 April 1914, 7 December 1914, 4 January 1915, 20 December 1916, 8 January 1917 and 5 March 1917.

[100] Ibid., 4 November 1918, RCPSG, 1/1/1/14.

[101] Ibid., 4 February 1918.

[102] Ibid., 13 January 1919.

[103] William Robertson Snodgrass (1890–1955). He later joined the venereal service in Glasgow. He remained also at the GWI, where he later served as sub-chief to George Allan, where he developed an interest in cardiology. He later became the first physician appointed to the Southern General and was instrumental in transforming the old Poor Law institution into a first-rate general hospital. He built up a large consulting practice in general medicine and, in

to the outbreak in a letter read at the 12 July Faculty meeting. He suggested meeting with the City Council and producing posters and newspaper adverts urging vaccination and revaccination. In June the Faculty authorised him and Chalmers to draw up a 'reasoned pronouncement' of the argument for vaccination for the papers, while the President convened a meeting of the Glasgow eastern branch of the BMA to mobilise 248 practitioners to take part in a house-to-house vaccination service. The newly formed Scottish Board of Health was to be asked to provide these practitioners with supplies of free lymph for every insured person, and to announce that vaccination was free to the insured.[104]

The Faculty had passed little comment on the establishment in 1919 of the Ministry of Health or Scottish Board of Health (except to complain about the medical representation on the latter).[105] It merely approved of the other notable events in health policy in these years, the Child Welfare Act of 1918 and the Nurses (Registration) Act of 1919, although it did send a communication to the House of Lords, the Prime Minister and the Secretary of State for Scotland urging that the latter should include, a 'single standard of training, a uniform examination, and a single register', so that Scottish nurses would not have to seek separate entry to an English register before practising south of the border.[106] Again the Faculty was only concerned about the establishment of the Scottish Consultative Council on Medical and Allied Services in 1920 in respect of Faculty representation on it.[107] However, this body's emphasis on the central role of the family GP in any national medical service,[108] a view echoed in the Dawson (1920) and Cathcart (1936) Reports, may have contributed to the Faculty's lack of concern for the Fellowship in these years. If the coming national medical service was to be based on the extension of the NHI scheme to provide GP cover for the whole population, with the hospital service remaining largely untouched, there would not be tremendous pressure for the reform of postgraduate examinations for a new

---

1942, was appointed consultant physician in charge of wards (i.e. a 'chief' in his own right). He was the first to use penicillin at the GWI and took part in trials of the sulphonamide therapy for VD. He was a Fellow of Faculty, Honorary Librarian from 1933–46, Visitor in 1947 and President from 1948–50. In the latter capacity he represented the Faculty on the Joint Consultants Committee in its negotiations with the government about the form of the NHS, and he had attempted to organise a Scottish consultants committee before the Second World War. Obituaries, *BMJ*, 3 December 1955; *Lancet*, 3 December 1955.

[104] *Faculty Minutes*, 12 April and 7 June 1920.
[105] Ibid., 7 April 1919 to 7 June 1920, passim.
[106] Ibid., 16 December 1919.
[107] *Faculty Minutes*, 12 April, 3 May and 7 June 1920.
[108] See Craig, *History RCPEd*, p. 711.

generation of hospital consultants. This notwithstanding, the financial difficulties that St Mungo's Medical School was experiencing during these years,
due to lack of students, might have alerted the Faculty to its own dwindling
core educational role. In 1916 the Council responded to these difficulties by
issuing a report on the proposed merger of St Mungo's and Anderson's, or
the complete closure of St Mungo's, which advocated the preservation of
the status quo. The rising importance of specialisation in medicine might
also have prompted the Faculty to reform the Fellowship into a more modern,
relevant examination. Such future problems were, however, temporarily
disguised by the flood of post-war candidates for the Fellowship, noted at
the outset of this chapter. Yet this was to prove only a short-lived respite
from the increasingly serious fundamental problem of the drift towards
university undergraduate examinations and the less than satisfactory reputation of the Fellowship.

Some faltering attempts were made in this period to expand the Faculty's
core licensing role by formulating new postgraduate diplomas to add to the
existing DPH. In 1904 a plan for a Diploma in Tropical Medicine and Hygiene
reached an advanced stage – the three Scottish corporations agreed on a
conjoint scheme for awarding the diploma together – before the RCPEd
scotched the plan, doubting the wisdom of competing with the university
equivalent.[109] In 1908–10 there were also thoughts of a Diploma in Psychiatry,
prompted by a request for a new examination from the Medico-Psychological
Association. Nothing came of this however, as the Faculty decided to wait for
the GMC to pronounce on the scope required for such a qualification. The
Faculty did not at this stage feel qualified to embark on the specialised
postgraduate *teaching* that would be required to prepare candidates for
such an examination.[110] However, one new diploma was launched, in
Ophthalmology, in 1921.[111]

Such diplomas were anyway aimed largely at GPs or MOsH, perhaps
because these were contemporaneously perceived to be the core workers
(and thus the potentially expanding market) in any future national health
service. There was, however, no formalised postgraduate teaching to prepare
candidates for the Fellowship, an examination aimed at those who wished
to become hospital consultants, with private specialist consulting practices.
Young practitioners had to rely on a series of junior appointments in the
hospitals to gain a rounded experience, reading journals, and attending
meetings of the medical societies (like the Pathological and Clinical Society
and the Medico-Chirurgical Society of Glasgow) which met in Faculty Hall.

[109] Ibid., 5 December 1904.
[110] Ibid., 6 December 1909.
[111] Ibid., 4 May 1921.

The first quarter of the twentieth century, nevertheless, saw the beginnings of organised postgraduate education in both Glasgow and Edinburgh. In Glasgow, the Faculty was not directly involved but had a presence through the involvement of the hospital clinical teachers, most of whom were Fellows of Faculty and held teaching positions at the extra-mural colleges.

For many years before the First World War various Glasgow hospitals had offered postgraduate classes in medicine, but there was no centrally correlated scheme. The first attempt to organise postgraduate medical training came in 1914 with an initiative from the University Principal Sir Donald MacAlister and the Dean of the Medical Faculty Noel Paton. A meeting of the teaching staffs of the University Medical Faculty, the extra-mural medical schools (St Mungo's and Anderson's), and the general and special hospitals of Glasgow was convened at the request of the Medical Faculty in March 1914. MacAlister presided and said that in an investigation into postgraduate teaching in the United Kingdom he had been struck by its 'sporadic and imperfectly organised character'. The Glasgow medical school was in competition in providing training and qualifications with Edinburgh and London and 'a more imposing appearance' would be made 'in what was becoming an international movement' by instituting regular rather than occasional courses. MacAlister stated that as well as providing an organised system of training for higher qualifications, any scheme should also aim to make Glasgow a centre for refresher courses for practitioners from the remoter parts of Scotland.[112] Before the outbreak of war postponed its activities a General Committee for Post Graduate Medical Teaching in Glasgow was formed with two representatives from the University Medical Faculty and one each from the extra-mural schools and the Glasgow hospitals. This body met twice (at least once at Faculty Hall), drew up a survey of all facilities for post-graduate teaching and planned to issue a syllabus in early 1915.[113]

The committee reconvened in February 1919. Its work was now given an added stimulus by the need for refresher courses (both clinical and systematic) for graduates who had taken their degrees during or shortly before the war and whose careers had been interrupted by military service.[114] The committee felt that graduates would now be reluctant to pursue post-graduate training in Germany and Austria, whose medical centres had attracted large numbers of graduates before the war. Furthermore, Edinburgh and London had already announced emergency post-graduate schemes to deal with the problem of

[112] *Minute Book of Glasgow General Committee for Post-Graduate Medical Teaching/ Glasgow Post-Graduate Medical Association*, GUABRC, Med 2/1, 6 March 1914 meeting.
[113] Ibid., meetings of 13 May, and 12 June 1914.
[114] Ibid., 19 February 1919.

demobbed practitioners.[115] But, as MacAlister had noted at a meeting of the University Court on 7 November 1918, it was important for Glasgow not to be left behind in this national innovation meant not only 'for war-worn men but for practitioners at home' for whom 'the need for brushing up from time to time their knowledge and of coming into contact with the newer methods in the laboratory was becoming very urgent'.[116] Dr William R. Jack[117] and Dr A. M. Kennedy[118] as secretaries liaised with the Medical Superintendents of the Glasgow Hospitals and drew up a list of facilities for teaching and a preliminary syllabus for a Glasgow emergency scheme. The committee felt that the teaching 'should be in the main practical and clinical, and that courses in the specialities or earlier subjects of the curriculum should be arranged for in accordance with the needs of the returning graduates'.[119] The teaching was to be organised under two schemes. Scheme A was an attempt to systematise post-graduate clinical experience. Under it five or six graduates would go to each participating clinic and each graduate would have charge of a number of beds and be 'responsible for the investigation and recording of the cases and for all special investigations in connection with them, working directly under and along with the physician or surgeon in charge'. Scheme B consisted of specialist classes offered by the teaching staffs of the various hospitals, either lectures or demonstrations.[120]

The first courses began in the summer session (May to June) of 1919. There were no fees for Scheme A (the committee felt that returning practitioners had already paid with their war service), but for Scheme B the charge was five guineas for the two months. This first course attracted forty-three graduates: one on Scheme A only; twenty-five on Scheme B only; and seventeen jointly on both.[121] A further course was run in September and October. In total eighty-six graduates attended during 1919; most were general practitioners returning from the services but a few were GPs who had been in

[115] 'Post-Graduate Medical Teaching: The Position in Glasgow', issued by the committee, GMJ, new series, 93 (1920), p. 218. See also Minute Book of GGCPGMT/GPGMA, 24 Febuary 1919.

[116] 'Current Topics: The University of Glasgow', GMJ, new series, 91 (1919), p. 46.

[117] Physician at the GRI, 1913–26 and Honorary Consulting Physician, 1926–27.

[118] The fiery red-haired Alexander Mills Kennedy graduated MB, ChB from Glasgow University in 1908. He became assistant to Robert Muir, the Professor of Pathology at the University, and then Senior Assistant Pathologist at the GRI (where he was clinical assistant to Professor Walter K. Hunter). He also worked as Pathologist to the Glasgow Maternity and Women's Hospital. He later became Professor of Medicine at Cardiff University. He wrote many pathological papers on the heart.

[119] Minute Book of GGCPGMT/GPGMA, copy of letter to Medical Superintendents of Glasgow Hospitals from General Committee, 26 February 1919.

[120] Ibid.

[121] Minute Book of GGCPGMT/GPGMA, 16 June 1919 meeting.

general practice throughout the war refreshing their knowledge. Classes were held in medicine, surgery, obstetrics and special subjects. In addition a special course in clinical and practical tuberculosis was arranged at the Consumption Sanatorium, Bridge of Weir, and at its Tuberculosis Dispensary in Glasgow. Special evening demonstrations were also arranged in diseases of the ear, nose and throat.[122]

In March 1920 a permanent scheme was approved by the committee and was advertised in the GMJ.[123] Under the new scheme the committee became the Glasgow Post-Graduate Medical Association. Like its predecessor, it was made up of representatives of the teaching institutions (the University Court and Medical Faculty and the managers of the leading hospitals) and of the teachers themselves. At the initial meeting of the new association, Leonard Findlay suggested that it should have a central office based at Faculty Hall,[124] but Sir Hector Clare Cameron successfully argued that, 'as the Post Graduate teaching was intended to appeal to graduates all over the world it would be better to have the office associated in some way with the University'.[125] In this period, then, the University wished to retain control of as much medical teaching as it could. Later, when the battle for control of undergraduate teaching was won, the University was content to delegate postgraduate responsibilities to the Faculty.

At a time when medical specialties were only beginning to develop and be formally recognised, the association was naturally unclear as to what all this training was for: this was a pioneering scheme in search of a recognised qualification. It was aimed at the general practitioner, but also at the emerging specialist who might be studying for a Fellowship of a London, Edinburgh or Glasgow medical corporation, although candidates for these mainly continued to utilise private courses. The 1920 advertisement also makes it clear that the University was considering setting up special diplomas in such subjects as ophthalmology, psychological medicine, radiology and electrology, dermatology, otology, laryngology and tropical medicine. In the short term such training was also thought to be relevant to the expanding roles

[122] GMJ, 'Post-Graduate Medical Teaching: The Position in Glasgow',
[123] Ibid.
[124] J. Leonard Findlay worked in the Pathological Department of the Western Infirmary under Robert Muir between 1905–12, and was at the same time assistant to Professor Samson Gemmell in the hospital. In 1914 he was appointed physician to the newly built Royal Hospital for Sick Children, and became Leonard Gow Lecturer in Medical Paediatrics in 1919, progressing to the new Samson Gemmell Chair of Paediatrics in 1924. Here he undertook and encouraged research which attempted to relate laboratory and clinic. Obituary in GMJ, 28 (1947) pp. 232–34.
[125] Minute Book of GGCPGMT/GPGMA, meeting of 14 June 1920.

of Maternity and Infant Welfare Officers, Tuberculosis Officers and School Medical Officers, and hence it was planned to introduce advanced courses in obstetrics and child welfare; school medical inspection and hygiene; tuberculosis; and venereal diseases. By the 1921 session the managers of the GRI, GWI, Victoria Infirmary, the Royal Maternity and Women's Hospital, the Royal Samaritan Hospital, the Royal Hospital for Sick Children, the Eye Infirmary, the Ear, Nose and Throat Hospital, the Lock Hospital and the Royal Asylum (Gartnavel) had all agreed to provide facilities for weekly demonstrations without charge. The staff of these hospitals agreed to give demonstrations from 17 November through to 25 May 1922 on Wednesday afternoons from 4.15 p.m. for one to one and a half hours. The fee of £3 3s. 0d. for the whole course was payable not to the hospitals, but to the association for the furtherance of postgraduate education in Glasgow.[126] Forty-nine graduates enrolled in 1921, five as clinical assistants, and the remainder in the clinical courses.[127]

This association scheme continued throughout the 1920s and 1930s offering graduates clinical assistantships in the hospitals (for a minimum of three months), advanced full-time courses of six months' duration; and weekly demonstrations on Wednesday afternoons for local practitioners. The Department of Health for Scotland gradually began to send practitioners from isolated Highland areas to the full-time courses – paying the fees and arranging for a locum. By 1938 this had developed into an enlarged scheme by which the department attempted to ensure that all National Insurance practitioners should be posted to courses in one of the Scottish teaching centres every five years at the Department's expense. In 1938 ninety-three of the 141 graduates studying in Glasgow belonged to his scheme.[128]

Thus postgraduate teaching began in Glasgow under University auspices, but such activities were to become centred on the Faculty after the inception of the NHS, as we will later observe.

The Fellowship

Despite this scheme's lack of training for higher qualifications in medicine and surgery, its usefulness for returning practitioners may have influenced the temporary revival that the Faculty's Fellowship experienced in the immediate postwar years, due to the pent up demand of doctors serving abroad. As Table 2 shows the numbers of Fellows admitted rose from an

[126] Ibid., 10 November 1920 meeting.
[127] Ibid., 20 December 1921 meeting.
[128] 'The Glasgow Post-Graduate Medical Association', GMJ, 27 September 1946, pp. 249–52, one in a series of articles on 'Medical Education in Glasgow'.

average of around five per year to twelve in 1918; twenty-six in 1920; dropping back to nine in 1930.[129]

The arrangements for the Fellowship underwent a number of changes during this period. A positive development was the admission of women to the Fellowship. Faculty opposition finally ended in 1912. The first application by a woman to become a Fellow came in March 1897 from Elizabeth Adelaide Baker, a registered medical practitioner. On taking legal opinion, the Faculty was advised that it was not entitled to admit women as Fellows according to its charter, but this would not have been an unpopular conclusion within Faculty.[130] In July 1905, Dr Jessie M. MacGregor, who asked again on behalf of a conference of medical women practising in Glasgow, was again met with a firm masculine no from the Faculty. At last, in 1912, the Faculty relented, and the first female Fellow, one Yamani Sen, a Licentiate in Medicine and Surgery of the University of Calcutta was admitted as Fellow *qua* Surgeon by examination on 3 June 1912.[131] The Faculty must have remained a rather unwelcoming place for female Licentiates and Fellows for many years, however, as the first women's toilet was not installed until the 1950s.

In 1890 the original three modes of admission (the normal two-part examination for younger practitioners; the revised scheme of medicine/surgery plus one optional subject for practitioners qualified for six years; the small number of admissions by virtue of distinction in 'original investigations' as evinced by published works; and *ad eundem* admissions for members of other corporations) was streamlined. The Fellowship was now modelled entirely on that of the Edinburgh Surgeons (though it was not that much different from that of the Edinburgh Physicians). The first method was now abolished and admission was now to be largely by the second, that is an examination in medicine/surgery and one optional subject,[132] though the candidate now had only to be qualified for two years. The other two forms of admission were left untouched except that Fellows admitted on the basis of professional distinction could now be nominated for their 'scientific attainments or eminence in Medical practice'.[133] Four such Fellows could now be admitted each year (this was up from two in 1885), and this when

---

[129] See *Register of Fellows, 1892–1962*, RCPSG, 1/4/9/1.

[130] *Faculty Minutes*, 1 March 1897.

[131] Ibid., 2 September 1912.

[132] Medicine continued to include clinical medicine, medical pathology and therapeutics; and surgery to include clinical surgery, operative surgery, surgical anatomy and surgical pathology. The optional subjects were anatomy, physiology, pathology, midwifery, diseases of women, medical jurisprudence, ophthalmic surgery, aural, laryngeal and nasal surgery, dental surgery, state medicine, psychological medicine and dermatology.

[133] *Faculty Minutes*, 2 June 1890, RCPSG, 1/1/1/11.

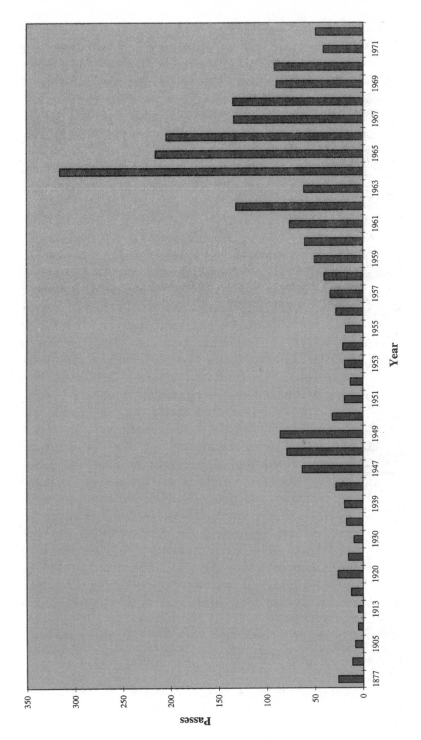

Table 2. *RCPSG: Numbers Passing Fellowship/Membership in Medicine, and Fellowship in Surgery, 1877–1972. (Source: RCPSG, 1/4/9/2, 1/4/10/2A, 1/4/11/1, 1/4/12/2)*

there might only be five Fellows admitted by examination during a year.[134] The fee was £25, rising to £30 in 1894, but £50 if the Fellow wished to qualify to hold Faculty Office. These changes of 1890 made the Fellowship an easier examination, covering a more limited set of subjects, and allowed admission without examination for an increased number per year for clinical as well as scientific expertise. The changes, as recommended by the Council, were carried by a narrow majority of seventeen to fourteen, with nine Fellows unusually registering protests.[135]

The First World War saw a further easing of the conditions for admission to the Fellowship. At the Faculty meeting of 1 November 1915 it was decided that the regulations should be altered to allow entrance to the Fellowship to those practitioners who had been on war service after passing an examination simply in one subject from Pathology, Bacteriology, Midwifery and Diseases of Women, Ophthalmology, Aural, Laryngeal and Nasal Surgery, State Medicine, Psychological Medicine, or Dental Surgery. This policy was to continue for five years from the end of the war,[136] but was still running in 1924. The candidate's wartime medical experience had to be in the chosen area, and the Council would decide whether the experience was sufficient before the candidate was able to proceed to the examination. This scheme was perhaps perceived by Faculty as a patriotic gesture. A notice was inserted in medical journals and the King was to be informed.[137] It may also have been intended to boost the numbers presenting themselves for the Fellowship examination, which it certainly did, although this upturn could equally well have been caused by a build up of demand during the war years. The special scheme was much easier than the normal route to the Fellowship, which still consisted of examinations in medicine or surgery *plus* one in a subject chosen from a list which had, by 1924, taken account of specialisation and included: anatomy; physiology, pathology, bacteriology, midwifery, gynaecology, medical jurisprudence, state medicine, specific fevers, dermatology, psychiatry and neurology, dental surgery, ophthalmology, aural, laryngeal and nasal surgery, diseases of children, and venereal diseases and their sequelae.[138]

In 1934, while the number of Fellows that could be admitted on the basis of distinguished scientific or clinical work was decreased from 1924's three to two, the terms on which such admissions (which had to be approved by

---

[134] 'Regulations of the Faculty of Physicians and Surgeons of Glasgow', November 1894 (Glasgow, 1894), RCPSG, 1/5/1/9.

[135] *Faculty Minutes*, 7 July 1890.

[136] *Faculty Minutes*, 1 November 1915.

[137] Ibid. The Secretary of State for Scotland was informed, but replied that he did not think it worth informing the King via the Privy Council.

[138] 1924 details from 'Regulations of the RFPSG', 1924, RCPSG, 1/5/1/11.

a Council report to Faculty) were dramatically broadened. Now the individuals might be:

> either Professors in a Medical Faculty of a University or of a Medical School recognised by the Council of Faculty, or Senior Medical Practitioners in the Public Services, or Senior Medical Practitioners engaged in clinical or other hospital work involving instruction in the hospital of students of a University or of a Medical School recognised by the Council of Faculty, or in research work.[139]

Fellows thus admitted did not now even have to be qualified for at least twenty years nor of at least forty-five years of age, as had previously been the case. The motion thus to amend the regulations had still been passed in spite of the written intervention of one Dr E. Gordon Mackie, a Fellow resident in Sheffield. Mackie had written that he felt

> strongly that the Council ought to be made aware of the fact that while the Fellowship is not well known in England, in such Medical circles as know it, it is regarded as a semi-honorary distinction obtained by election somewhat in the manner of the Fellowship of the Royal Society of Medicine, and not as implying status of the type ensured by examination as is commonly associated with the Edinburgh Fellowship or Membership. The present trend of widening the grounds for election without examination, can only justify instead of falsifying this estimate, and must inevitably reduce the value of the honour it is sought to confer.[140]

An anecdote told by Stanley Alstead confirms the poor reputation of the Glasgow Fellowship in the interwar period. When Alstead was Pollok Lecturer in Materia Medica in the University of Glasgow (1932–36), his clinical chief in the Western Infirmary, where Alstead was also Dispensary Surgeon, was Dr James Carslaw.[141] Alstead commented that, 'he [Carslaw] literally forbade his junior colleagues to sit the Fellowship examination of the Faculty or indeed "to have anything to do with that place in St Vincent Street"'.[142] We will see how the Faculty finally set about remedying the problem of the Fellowship, once it had become essential to the institution's survival. Ironically, as we shall see, if Stanley Alstead had taken Carslaw's advice, the ultimately successful outcome might not have been achieved.

---

[139] Faculty Minutes, 2 July 1934.

[140] Ibid., letter from Dr E. Gordon Mackie to the Secretary of the RFPSG, 30 June 1934.

[141] Carslaw was one of the Visiting Physicians. He had graduated with a Glasgow medical degree in 1892, and had been House Physician to Sir William Tennant Gairdner (as had two of the three other Visiting Physicians, Barclay Ness, T. K. Munro – the other was Stockman. Carslaw, Ness and Munro were all ex-residents of the GWI and had served through its junior staff grades). He was known as 'the Doctors' Doctor' because of the popularity of his services with other practitioners when they were ill. See Loudon and MacQueen, The Western Infirmary, pp. 75–76.

[142] Alstead, Extracts, p. 508.

3

Reorientation and Revival, 1930–1950

Two World Wars and the intervening phase of irresolution seemed to have made no impact upon its affairs and it is only since 1948 that renewed interest and growing enthusiasm have stimulated a renaissance. For the historian of the future it would be helpful to trace the seeds of this present renaissance and to record the names of those who contributed to it, but such a task would be impossible as well as invidious at this time. In part, the rebirth must be attributed to a few individuals and especially to the devotion of a small group of office-bearers and councillors, but in large measure it has been simply a reflection of the increasing activity of the Glasgow Medical School as a whole.[1]

The Faculty was an ailing institution in the early 1930s. Its qualifications and examinations were locked into the final stage of a long decline. The TQ was under increasing pressure from the undergraduate university degree, and the Fellowship was too open and not perceived by the profession as of a high enough standard to act as a passport to senior hospital positions. However, the increasing specialisation and scientisation of medicine, the way in which these changes came to Glasgow, and the enactment of a new government health policy which institutionalised these changes, were all to shake up dramatically the role of this ancient, but largely moribund, medical authority.

Medicine had become more specialised, and the work of hospital units was coming to reflect this change. The introduction of the NHS as a hospital and consultant centred system of health care provision both reflected and formalised these changes in the nature of medicine. Levels of consultant cover were now set in different specialties and a whole new staffing structure was developed to underpin the work of the consultants and to provide a training ladder up which junior hospital practitioners could progress to the consultant grade. Increased specialisation meant that this structure had to be flexible so that the intending consultant was able to gain practical experience of as many aspects of medicine and surgery as possible. Medical graduates were very keen to secure a foothold on the training ladder, because NHS posts were full-time (although the junior posts were of limited duration)

[1] Illingworth, *RCPSG: Short History*, p. 11.

and this meant an end to practitioners having to assemble a portfolio of medical jobs to earn a medical living.

After the Second World War there was thus intense competition for hospital posts, and a desperate need for organised training to complement and supplement the clinical experience provided by the training ladder. There was also a need for higher qualifications which would act as a passport to consultant status to be synchronised with the training ladder. At this point the local organisation of postgraduate education became an absolute necessity. The universities had now come to dominate undergraduate medical education, and in Glasgow the extra-mural schools were absorbed in 1947. This left the Faculty with no role in undergraduate education and a post-graduate qualification of dubious status unsupported by organised training programmes. However, the Faculty, in conjunction with the University had began to rethink its role while these changes were perceptibly gathering momentum in the late 1930s and early 1940s.

In general the national movement towards a national health service acted as a stimulus to the Faculty and enabled it to determine a socially valid and economically viable new role, within the new organisational structures and educational priorities of the service. The key influential documents which enunciated this change were the Goodenough Report[2] and, in Scotland, the Hetherington Report.[3] However, both of these accentuated and formalised developments in the education and organisation of the medical profession which had been maturing since at least the end of the previous century. Some of the educational antecedents have already been noted in the nineteenth-century reform debates. These are inextricably linked to the rise of medical and surgical specialties, the growing role of the state in the provision of medical services, and, equally importantly, to informal, intra-professional debates and agreements. The same is true of organisational changes in the provision of medical care. A combination of factors had led to the changes of organisation in the National Health Insurance (NHI) scheme of 1911, and ever since then there had been an ongoing debate in the profession and in government about the mechanics of further expansion of the state system. The achievement of the wartime reformers was to integrate the educational and organisational aspects of the problem into a compre-hensive and coherent solution. The opportunity was provided by the national

[2] Ministry of Health, Department of Health for Scotland, *The Report of the Inter-Depart-mental Committee on Medical Schools* (London, 1944). Known by the name of its chairman, Sir William Goodenough.

[3] Department of Health for Scotland, *Report of the Committee on Post-War Hospital Prob-lems*, Cmnd 6472 (Edinburgh, 1943).

concern with reconstruction, and the job was completed (despite Bevan's considerable problems with a medical profession understandably concerned with questions of autonomy now the crunch had come) during the following period of relative political consensus (on ends, if not means).[4]

The new integrated system of hospital organisation and medical education was centred on the local university. The university degree finally became the effective single portal of first entry to the profession, and by 1953 a pre-registration year of hospital experience had been introduced for the newly-qualified medical student. The new educational arrangements were synchronised with hospital staffing policy. The division of the profession into GPs and consultant specialists (which had been crystallising with the development of specialties since the late nineteenth century and was both acknowledged and stimulated by the informal adoption of the referral system by the profession from the early 1900s) was set in stone in the post-war staffing arrangements. After the initial qualification, doctors could either opt for a career in general practice or for a hospital career. If the latter, they began to progress through a hierarchical structure of training grades towards consultant status. This development meant that the higher education of the potential consultant was now the crucial educational issue. The Goodenough Report recommended that the royal medical corporations should have responsibility for such postgraduate qualifications. Their Fellowships would become (either formally or informally) the passports to consultant status. The London and Edinburgh Colleges began organising themselves on this basis during the war, and in Glasgow the Faculty redoubled its efforts to reestablish the Fellowship now the way ahead was clear.

However, it was the University which was the local linch-pin of the new integrated system of education and organisation. It had complete responsibility for undergraduate medical education (including control of earmarked University Grants Commission -UGC – funds to the hospitals for clinical teaching, and effective control of the majority of teaching posts), and command of UGC funds for the provision of the taught elements of postgraduate education. The University thus had a strong financial leverage in the organisation of the hospitals and this was reinforced by its strong representation on the new Regional Hospital Board (RHB), and on the Board's Medical

---

[4] On the coming of the NHS, and the medical reception to government plans see, *inter alia*, Charles Webster, *The Health Services since the War*, i, *Problems of Health Care: The National Health Service before 1957* (London, 1988); Michael Foot, *Aneurin Bevan*, ii (London, 1973), chapters 1–4; Daniel Fox, *Health Policies, Health Politics: The British and American Experience, 1911–1965* (Princeton, 1986); Honigsbaum, *Division in British Medicine* and *Health, Happiness and Security: The Creation of the National Health Service* (London, 1989); and James S. Ross, *The National Health Service in Great Britain* (London, 1952).

Education Committee (which arranged undergraduate education) on the Board's Appointments Committee, and on the Glasgow Postgraduate Medical Education Committee (GPGMEC), which, peculiarly to Glasgow, continued to organise postgraduate education after the Appointed Day – 5 July 1948 – when the new NHS came into being.[5] In Scotland this influence was increased as the teaching hospitals were included in the regional organisation and were responsible directly to the Regional Board.[6] In Glasgow this influence was even more pronounced because many of the key players in the hospital clinical elite had strong University affiliations. This, in turn was not accidental, but was partly due to a deliberate policy of medical empire-building pursued by the University under two Principals in particular: the eminent medical man Sir Donald MacAlister (Principal 1907–29), and especially Sir Hector Hetherington (Principal, and Vice-Chancellor 1936–61). It was MacAlister who had presided over the University's return to clinical teaching at the GRI. Hetherington rebuilt the University Medical School, introducing the modern academic concept of medicine, founded on clinical teaching and research using laboratory methods in specialised units, into the hospitals via a series of key full-time professorial appointments to academic clinical units. This process was begun before the inception of the NHS but was attuned to the later reforms. The academic units were to function as research schools radiating modern scientific, specialised, research-based medicine. The new Medical School needed local high-quality postgraduate training to supply staff for the hospitals. Since the University would be busy with its monopoly of undergraduate medical education in the post-NHS period, another separate but closely integrated body was needed to organise postgraduate matters. This was the role that Hetherington allocated to the Faculty. Prescient minds in the Faculty Council realised that this was an opportunity that their ailing institution desperately needed to survive in the post-NHS medical world, and they set about placing the Faculty on its new trajectory.

The Faculty story is thus intimately linked with the story of the University Medical Faculty in this period. Former disagreements melted away, not just because of the new realities of University dominance in medical matters, but also because the key Faculty players were often also key University men. This in itself of course reflects the new dominance of the University, but it should not be overstated. The Faculty continued to represent the Glasgow clinical

[5] Elsewhere this role was performed by a sub-committee of the Medical Education Committee.

[6] In England, where the medical schools were based on the great teaching hospitals and not the universities, these hospitals were independent of the Regional Board, and were accountable directly to the Minister of Health.

elite on the regional board and in the GPGMEC. There were differences of emphasis and of principle in debates on the content of medical education (both undergraduate and postgraduate). Broadly speaking, the University (particularly under Hetherington, and George Wishart [7] as Dean of the Medical Faculty and Director of Postgraduate Medical Education) tended to emphasise the importance of laboratory methods and specialisation (in teaching and research), while the Faculty, while not denying the importance of science or specialisation in modern medicine, were keen to push the traditional Glasgow emphasis on general clinical experience.

Charles Illingworth was in a very privileged position when he wrote the quotation which begins this chapter. He was one of a number of new key appointments made to the University Medical Faculty in the mid to late 1930s. These men were charged with bringing Glasgow medicine methodologically up-to-date, fostering in the hospitals a medical culture of specialisation, science, and clinical research as well as patient care. All of them were also key players in reforming the Faculty into a postgraduate teaching and examining arm of the Glasgow Medical School, providing an educational structure to correlate with the new hospital organisation. As Illingworth has argued, the destinies of the Faculty and the University were inextricably linked in this period. The increasing influence of the University and academic medicine based on specialisation and clinical research was aided by the birth of the NHS, which institutionalised the local hegemony of the University Medical Faculty. This conjunction enabled key players in University and Royal Faculty to devise together a crucial new role for the ancient medical corporation in the new medical scene. In this period the Royal Faculty took its first steps towards a new incarnation as a teaching and examining body in post-graduate medical qualifications.[8]

However in the late 1930s the Faculty's future was far from clear. Illingworth said of the Faculty in this period that, 'it must be confessed that in 1939 for many years it had fallen on evil days'.[9]

[7] George McFeat Wishart (1895–1958), Grieve Lecturer in Physiological Chemistry in the University of Glasgow 1921–35, Gardiner Professor of Physiological Chemistry 1935–47, Dean of the Faculty of Medicine and Director of Postgraduate Medical Education, 1947–58.

[8] The most useful of the limited accounts include Tom Gibson, *RCPSG*; Gavin Shaw, 'The Royal College of Physicians and Surgeons of Glasgow', in Derek Dow and Kenneth Calman (eds) *The Royal Medico-Chirurgical Society of Glasgow: A History, 1814–1989*, pp. 102–6 (Glasgow, 1989); Sir Charles Illingworth's autobiography, *There is a History in All Men's Lives* (Glasgow, 1988), passim. Sir Charles Frederick William Illingworth (1899–1991) graduated in medicine from Edinburgh in 1922, later taking the degrees of ChM and MD. He became FRCSEd in 1925. In 1939 he became Regius Professor of Surgery at Glasgow with wards in he GWI. He was RCPSG President 1962–64.

[9] Illingworth, *All Men's Lives*, p. 112.

Problems, 1938–41

In the first third of the twentieth century the Faculty functioned as the provider of registrable alternative undergraduate qualifications (the TQ, the Licence in Dental Surgery and the Higher Dental Diploma). Most candidates were students from the extra-mural colleges of medicine (Anderson's and St Mungo's) who, either for financial reasons or lack of relevant qualifications, did not wish to take a medical degree, but still wished to become medical practitioners. However, the Faculty's reliance on undergraduate examination was to prove a critical weakness. The numbers of extra-mural students began to decline after 1901 with the establishment of the Carnegie Trust with £2 million pounds of funds to pay the fees of Scottish university students. This decline was exacerbated by the introduction of a quota system to control the number and place of study of medical students in Glasgow from 1939.

The University, which was tightening its already strong grip on undergraduate medical education, and especially its new Principal, Hector Hetherington, was the main instigator of this scheme, which had a direct impact on the Faculty, as the minutes record, 'In view of the close connection between the welfare of the extra-mural schools and the Faculty as a Licensing Body'.[10]

The proximate cause of the dispute was the admission of large numbers of Jewish American students to the extra-mural schools, where they studied for the Scottish TQ. These students had been effectively barred from their home Medical Schools by the operation of secret quotas on the number of Jewish students. Their exodus to Scottish schools made Scotland 'the overseas centre for the training of American born physicians during the 1930s'.[11] It also helped to keep the Faculty financially afloat in a period when home students were opting more and more for a University degree. During the 1936–37 session St Mungo's, the smaller of the two extra-mural colleges, took about one hundred Americans, while the numbers admitted to Anderson's reached a peak of 250 in 1938.[12] In the period from 1933–1940, 148 of these Jewish American students studied in Glasgow and took the TQ. This bred resentment among the honorary clinical staff at the University over the amount that extra-mural teachers were receiving in class fees. There was also concern in the Medical Faculty over whether the number of foreign students

[10] *Faculty Minutes* , 23 January 1939, RCPSG, 1/1/1/16.
[11] Kenneth Collins, *Go and Learn: The International Story of Jews and Medicine in Scotland* (Aberdeen, 1988), p. 120.
[12] Ibid., p. 117.

was adversely affecting the education of home students, given the finite facilities for clinical teaching in Glasgow (i.e. beds and teachers). Matters came to a head over the teaching of radiology by University lecturers to extra-mural students in the Pathology Department of the Royal Infirmary. In early 1938 the University was trying to discourage extra-mural students by making them enrol, pay class fees and even a form of matriculation fee as if they were University students, in order to take these classes. The President of the Faculty, John Henderson (who had recently been appointed Honorary Consulting Physician to the GRI, and was on the teaching staff of St Mungo's College in Medicine),[13] wrote to the GRI Managers asking them to reassure him that the traditional policy of not granting priority access to the University was still operating, and reminding them that when the University Chair of Pathology was founded at the GRI in 1911 St Mungo's had agreed to contribute £600 annually towards the Professor's salary, hence its title of the St Mungo (Notman) Chair.[14] As Henderson argued, 'students attending the Infirmary are essentially Infirmary students and should be treated as equal irrespective of the body whose qualification they intend to take'.[15]

A satisfactory arrangement was reached with the University over the numbers of extra-mural students taking pathology generally at the GRI by the end of 1938. The University admitted the claim of St Mungo's students, but still would only allow Anderson's students free access once all St Mungo's students had been accommodated.[16] In January 1939 the University Court sent a memorandum to the Faculty and the Deans of the two Schools urging that a definite working limit or quota be set on medical students in Glasgow. It argued that, especially given the prospect of war and the need to set aside emergency beds, and the revised University medical curriculum, which would mean an increase of around 50 per cent in demand for clinical teaching, a cap of 200 students per year should be set. The University explicitly stated that it was no longer willing to see 'this inflow of aliens ... continue' when the spaces could be filled by home students at the University. The memorandum put forward two schemes, one of voluntary and one of fixed quotas, both ensuring priority for home students. The alternative to acceptance of

[13] Henderson was originally appointed Physician to the Infirmary in 1913. See Jenkinson, Moss and Russell, *The Royal*, p. 279; and *Medical Directory* (1945).

[14] See Edith MacAlister, *Sir Donald MacAlister of Tarbert* (London, 1935), p. 329; Dupree, 'Development of Medical Education', pp. 22–23.

[15] *Faculty Minutes*, 7 February 1938, letter from President to the Managers of the GRI, RCPSG, 1/1/1/16.

[16] Sir Hector Hetherington Papers, GUABRC, DC8/995, 'Notes on a Meeting Held within Mr Bruce Warren's Office, 45 West George Street, on Tuesday 29 November 1938'.

one of these schemes was arbitration by the Department of Health for Scotland.[17]

In reply, a memorandum from a joint committee of the Faculty and the two schools disputed the University's assessment of the problem. University admissions had not been limited to 180 per year as claimed but had consistently exceeded that figure. Moreover, a number of clinical classes in the teaching hospitals were not filled. The memorandum argued that rather than a dearth of facilities for clinical teaching, 'there exists in the city a large reserve of clinical material which could be utilised for teaching purposes'. The Faculty memorandum claimed that the suggested quota systems would 'mean the disappearance of these Colleges which derive their revenue in large part from the fees of students and which do not receive Government assistance'.[18] The clear implication was that the University was trying to squash the extra-murals and gain complete control of undergraduate medical teaching in Glasgow.

The question of the number of medical students was settled by a temporary compromise: the governors of Anderson's and St Mungo's agreeing not to admit American students for the academic year 1939/40 so as to reduce the overall admissions. This satisfied the University. The following year a standing committee of the Faculty and the two schools was established to deal with all admissions to the extra-murals. The war had intervened to curtail the influx of Americans, so all parties were able to agree with the Faculty Council's recommendations that a quota be applied from October 1940. The University was to take 180 students and the extra-murals sixty between them (a figure more like their pre-American influx admissions), with a further ten places unallocated.[19]

This dispute marked the beginning of the end for the extra-mural schools. The new University Principal, Hetherington, clearly planned for the University to take over all medical teaching in the city. To do this he had to eliminate the extra-mural schools, and this was his first move against them. He was able to absorb the schools into the University in 1947, just as the National Health Service Act was about to enshrine the local importance of the University as the centre of medical education and the hub of the new service. There

---

[17] 'Memorandum by the University Court on the Admission of Medical Students to Glasgow', appended to *Faculty Minutes*, 23 January 1939, RCPSG, 1/1/1/16.

[18] 'Proposed Reply to the Memorandum from the University with Regard to Medical Teaching in Glasgow', appended to *Faculty Minutes*, 12 May 1939, RCPSG, 1/1/1/16. The final form of this memorandum differed only in the deletion of the last sentence. None of the above points or quotations were changed, see ibid., 5 June 1939.

[19] See ibid., 3 July 1939, 1 July 1940.

were early danger signals here for the Faculty, which relied on the extra-mural schools for candidates for the TQ.

As well as problems with the status of the Faculty as an undergraduate examining body, there were also problems with the main higher qualification of the Faculty: the Fellowship. By the 1930s the Fellowship had fallen into disrepute, being no longer considered an adequate qualification for senior appointments to hospitals (that is appointments as Honorary Physician or Surgeon, the equivalent of the modern consultant) anywhere in England, in Edinburgh and even in the Glasgow area. As already noted, examination standards had been lowered immediately after the First World War in an attempt to boost the low numbers of those attempting to become Fellows, and the negative impression of the Fellowship persisted. This was also exacerbated by the increasingly broad conditions under which Fellows could be admitted on the basis of their professional distinction without examination. The Faculty *seemed* to be a moribund professional club, rather than a stringent standard-setting body. Most candidates would take the more prestigious London or Edinburgh qualifications. The attraction of becoming a Fellow was thus limited to the opportunity to earn fees from becoming an examiner for the TQ or the Fellowship. In 1939, however, the perilous financial state of the Faculty forced the Treasurer to ask examiners to forgo such fees! [20]

There had been a move in 1936–37 to establish a formal course of training specifically for the Fellowship under the auspices of the Glasgow Postgraduate Medical Association. Two Glasgow surgeons and Fellows of Faculty, Archibald B. Kerr (then an Out-Patient Surgeon at the GWI and an assistant in Roy Frew Young's unit),[21] and W. Arthur Mackey (then a Resident with

[20] Illingworth, *All Men's Lives*, p. 112.

[21] Kerr had spent three years with Robert Muir, the Professor of Pathology as his assistant, and Assistant Pathologist to the GWI (1931–33), before his appointment as Dispensary Surgeon in Roy Frew Young's Surgical Unit. In 1945 he was sub-chief to Scouler Buchanan in the GWI and in 1946 he also succeeded George Dalziel as chief at the Royal Alexandra Infirmary, Paisley. Here he was also Honorary University Lecturer in Clinical Surgery from 1946–72, and it was during his time that Paisley (like the Southern General) came into the University orbit and became a teaching hospital. He was President of the Royal Medico-Chirurgical Society of Glasgow in 1951. He succeeded to his own wards at the GWI in 1954, where worked on surgery of the colon and rectum. This interest originated in his wartime medical experiences in the Middle East where he saw a lot of dysentery. He found his specialty socially embarrassing and when asked what he did would answer, 'Well, I'm really a general surgeon but I have a special interest in the alimentary tract and the further down you go the more interested I become'. He was Visitor from 1962–64, and President from 1964–66, also serving as a member of the Western Regional Hospital Board and on the University Court. See Gibson, *RCPSG*, pp. 272–73; MacQueen and Kerr, *The Western Infirmary*, pp. 147–48; and A. B. Kerr, interview with Peter MacKenzie, RCPSG, 18/14.

Archibald Young, the Regius Professor of Surgery at the Western Infirmary, with an additional appointment as Surgeon to the Southern General),[22] had been running a private seven-week course in surgery and surgical pathology for two hours a day from 24 February 1936 for a fee of ten guineas.[23] In early 1937 the two Faculty Fellows sent a memorandum to the Glasgow Postgraduate Medical Association (GPGMA) requesting its support in arranging a more comprehensive scheme of training for the Fellowship. They wrote that:

> It is clear that a Medical Centre, with the teachers, population and traditions of Glasgow, *should* be able to offer post-graduate training second to none. That an active post-graduate school is desirable cannot be disputed. The fame and academic prosperity of the medical schools of London, Edinburgh and Vienna ... depend considerably upon their reputations as centres of post-graduate instruction.
>
> The principal lectures and demonstration courses at present provided by the Post-Graduate Association admirably fulfil the requirements of the general practitioner who desires to increase his familiarity with modern methods. These courses however are not designed individually or comprehensively to meet the needs of the prospective candidate for higher qualifications, who requires a group of concurrent intensive courses appropriate to his special interests. We feel that such courses should be available in Glasgow in Medicine and Surgery alike ...[24]

What had sparked this appeal was probably advanced knowledge of the plans of the University of Edinburgh, in conjunction with the Royal Infirmary and various other hospitals, to run an eight-week intensive course on internal medicine in the autumn of 1937 under the auspices of the Edinburgh Post-Graduate Executive Committee. This Committee had previously, like its Glasgow equivalent, run only short refresher courses for GPs. The Edinburgh scheme included clinical attachments under the eight chiefs at the Royal Infirmary, systematic lectures by experts in the various Departments of Medicine, and discussion of clinical cases relevant to these lectures.[25] This

---

[22] Mackey continued to work at the Western until he was appointed to succeed J. A. G. 'Pop' Burton as St Mungo Professor of Surgery a the GRI in 1953. This, after Leslie Davis, was only the hospital's second full-time professorial appointment.

[23] Details from advertisement in *BMJ*, 10 March 1936, in GUABRC, Med 2/1.

[24] 'Memorandum to the Post Graduate Association Regarding the Provision of Courses of Instruction for those Contemplating Higher Examinations in Surgery – FRFPS Glasg., FRCSEd, FRCSEng, &c.' (1937), GUABRC, Med 2/1.

[25] See Edwin Bramwell (Chairman of the Edinburgh Post-Graduate Executive Committee), 'Foreword' to Bramwell (ed.), *Edinburgh Post-Graduate Lectures in Medicine*, i, 1938–39 (Edinburgh, 1940), pp. ix–xvi. Taking part in the Internal Medicine course launched in 1937 were all the physicians and surgeons of the Edinburgh Royal Infirmary, and all the staff of the Infirmary's Departments of Radiology, Venereal and Skin Diseases, plus those of the University

1 View of Tron Steeple

2 Faculty Hall in St Enoch Square

3 Faculty Hall in St Vincent Street

4 Black and White sketch of the College buildings in St Vincent Street, 1960s

5   Portrait of James Watson by Daniel Macnee

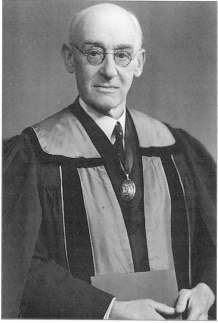

6 Roy Frew Young

7 Geoffrey Balmanno Fleming

8 The President, W. R. Snodgrass, laying a wreath on the tomb of Maister Peter Lowe after the Commemoration Service in the Cathedral on the 350th Anniversary of the Royal Faculty in November 1949

9 Portrait of Sir Hector Hetherington in Honorary Fellow's robes by Sir Stanley Cursiter

10 Archibald Goodall as a young man

11 Gavin Shaw as President in 1978 outside the College

12 Sir Charles Illingworth (1963) presenting the programme *Peptic Ulcer: A Surgeon Cross-Questioned* from the STV series 'Postgraduate Medicine' (*Scottish Media Group*)

13 Group shot (1963) from *Peptic Ulcer: A Surgeon Cross-Questioned* from STV series 'Postgraduate Medicine', left to right: Sir Charles Illingworth, Mr Gabriel Donald, Mr A. D. Roy (Senior Lecturer, Department of Surgery, GWI), Professor Bernard Lennox (Department of Pathology, GWI), Dr Gavin Shaw, Dr T. J. Thompson (Consultant Physician, Stobhill Hospital) (*Scottish Media Group*)

14 College Council at change of name (1962), left to right: (back) R. Barnes, K. Fraser, T. Symington, T. Anderson, T. Gibson, A. L. Goodall (Honorary Librarian), T. D. V. Lawrie, R. B. Wright (Honorary Treasurer), Mr Brown; (front) G. Shaw (Honorary Secretary), S. Alstead, J. H. Wright, C. F. W. Illingworth (President), A. Jacobs, A. B. Kerr (Visitor), D. F. Cappell

15 Professor T. C. White, the first Dental Convenor

16 College Past Presidents, July 1998, left to right: (standing) Professor Norman Mackay, Professor Arthur Kennedy, Dr Robert Hume, Sir Donald Campbell, Mr James McArthur, Sir Thomas Thomson; (sitting) Sir Ferguson Anderson, Professor Edward McGirr, Mr Colin MacKay (President), Sir Andrew Watt Kay, Dr Gavin Shaw

17 Sir Donald Campbell, PRCPSG, welcoming Diana Princess of Wales to the College on the occasion of the opening of the Princess of Wales Conference Suite, 8 June 1993

Edinburgh course continued for many years and its success was later to prove very influential in shaping the form of postgraduate teaching the Faculty was to offer in the 1950s. Another factor prompting the memorandum from the two Fellows was the recent development of postgraduate teaching in London, where, since 1935, there had been a specific hospital (the Hammersmith) dedicated to postgraduate teaching. The two Glasgow Fellows had been running their own private courses since 1934, they stated, but while they had 'received numerous replies' to their *BMJ* advertisement, the course had not proved particularly successful since these responses were:

> usually accompanied by enquiry regarding facilities for intensive teaching in clinical surgery, and a number of the specialties. Since this was not available the majority of the prospective students did not feel justified in coming to Glasgow for study. The provision of an adequate clinical courses, therefore is, in our opinion, the crux of the situation.[26]

The two surgeons argued that what Surgical postgraduates needed was not a clinical assistantship attached to one unit (which was currently on offer from the GPGMA), but rather experience of 'a succession of clinics' plus systematic instruction in Surgical Anatomy, Pathology and Operative Surgery, and, in some cases, a specialty.[27] They went on to outline a scheme of clinical and systematic study in which both the proposed surgical anatomy class and the clinical surgery sessions in the Glasgow hospitals were organised so as not to clash with University teaching times. It was suggested that the Anatomy class actually be held at the University. This degree of planning suggests insider knowledge, or even, conferral, with senior representatives of the University or its Medical Faculty. The GPGMA generally approved the memorandum in March, and by early May its Committee to Make Arrangements for Clinical Courses and Clinical Assistantships had resolved

Departments of Anatomy, Physiology, Pathology, Bacteriology and Tuberculosis, and those of the Royal Hospital for Sick Children, the Royal Hospital for Mental and Nervous Diseases, the City Fever Hospital and the Municipal Hospitals. Thirty-two graduates enrolled in 1937, and the same number in the autumn of 1938. See ibid., pp. xv–vxi. The Edinburgh Post-Graduate Executive Committee had begun organising GP courses in 1905. It was composed of three members each from the Medical Faculty and the Governing Board of the School of Medicine of the Royal Colleges (see ibid., p. xii). Its membership was thus not as broad as its 1914 Glasgow equivalent, as it did not include representatives of the hospital clinical teachers (see Chapter 2, above). For more details of how the Edinburgh Internal Medicine course in the late 1930s combined correlated clinical with systematic teaching see the entry on 'Edinburgh Postgraduate Courses', in the *BMJ* Educational Number, Session 1938–39, 3 September 1938, p. 529.

[26] 'Memorandum to Postgraduate Association' (1937).
[27] Ibid.

to sound out the chiefs of all the teaching hospitals. By September 1937 a meeting of fifteen surgical chiefs had been held and subjects suitable for clinical demonstrations had been discussed. The new Principal of the University, Sir Hector Hetherington, who had succeeded MacAlister in late 1936, had been interviewed and had provisionally offered the use of the Anatomy Department at the University, with the concurrence of the Professor.[28]

According to the minutes of the GPGMA which survive, these plans for a comprehensive course of postgraduate training for the Fellowship came to nothing in 1937. As the *BMJ*'s educational number reconfirmed in September 1938, the only formal postgraduate courses on offer in Glasgow under the GPGMA were either aimed at GPs or took the form of a clinical assistantship to one unit, with no rotation.[29]

The plans indicate a growing awareness of the importance of formal postgraduate training schemes to the health of medicine in Glasgow. Moreover, the identity of the participants was of great importance, considering later developments. First, the two Fellows who had suggested the scheme, after the failure of their own private courses at the Faculty, were linked with Archibald Young and Roy Frew Young. Archibald Young was then Chairman of the GPGMA, and had been in negotiations with the University to establish a modern Institute of Surgery at the Western Infirmary.[30] Roy Frew Young was to become a reforming President of the Faculty in 1940. His energies in this office were largely directed towards revivifying the Fellowship as a sound basis on which to build a new future for the Faculty given the trends in medicine and in national health policy. It is therefore of little surprise that we should find one of his protégés engaged in promoting a pioneering (if ultimately unsuccessful) scheme of postgraduate training as early as 1937. The cooperation of the new University Principal is also noteworthy. Hetherington's comprehensive plan for the reform of Glasgow medicine meant he had a strong interest in encouraging postgraduate training. When the 1937 scheme came to nothing, Hetherington soon became involved in a more ambitious scheme to harness the resources of the clinical teachers in the hospitals by helping to promote their natural home, the Faculty, as the local postgraduate teaching and examining centre.

---

[28] Details in this paragraph are from minutes of the meeting of the GPGMA, 4 March 1937, GUABRC Med 2/1; and GPGMA, Committee to Make Arrangements for Clinical Courses and Clinical Assistantships, Minutes, GUABRC Med 2/2, 5 May and 7 September 1937.

[29] *BMJ*, Educational Number, Session 1938–39, 3 September 1938, p. 529.

[30] This is suggested by uncatalogued private papers of Archibald Young recently (May 1998), loaned to the College by his son, the late Mr Stuart Young. A cursory reading of these papers is all that has been possible in the preparation of this book.

The attention of Faculty was further focused on the problem of the Fellowship in early 1939 when the Faculty received two long letters from a Fellow with a hospital appointment at the Liverpool General Hospital, Dr Samuel Saxon Barton,[31] which bewailed the lowly status of the Glasgow Fellowship. Saxon Barton informed Faculty that honorary staff appointed in the large Liverpool teaching hospitals could possess any of the higher qualifications on offer in the British Isles *except* the Faculty's. He suggested with heavy irony that at the next Faculty meeting a resolution should be proposed:

> To print in red letters on the back of the syllabus supplied to candidates who consider presenting themselves for the Fellowship of the Faculty either in Medicine or Surgery *Candidates are warned that due to the lack of energy and foresight displayed by the Royal Faculty of Physicians and Surgeons of Glasgow, this Fellowship is purely an academic qualification and will be entirely useless to them in the future should they attempt to earn their livelihood among the Barbarous English.*[32]

According to Saxon Barton the Fellowship was in similar disrepute, not only in Liverpool but also in the hospitals of Manchester and Leeds. English medical opinion of the Glasgow Fellowship did not seem to have changed since witnesses before the Royal Commission on the Medical Acts in 1881–82 had rehearsed apocryphal anecdotes about it being a substandard qualification taken only as a last resort by failed London or Edinburgh candidates.[33] Saxon Barton reported that it was still considered as: 'An examination which is the last hope of the second-rate candidate who has failed to get anything better, but who, for some reason, must attempt to get a higher qualification of some sort'.[34] He reported that he had canvassed medical opinion in Liverpool and that the consensus was that the root of the Fellowship's image problem was that it was not even valued 'at home' in Glasgow. Fellows often failed to list their Glasgow qualification on published work and, more importantly, it was not obligatory for candidates for surgical posts at the Royal, Western or Victoria Infirmaries to hold it. In his opinion if it was made obligatory for all physicians and surgeons applying for honorary posts at the Glasgow teaching hospitals, the Fellowship would 'at once be thought more of South

[31] Samuel Saxon Barton (1890–1957) An obstetrician and gynaecologist. After medical education at Liverpool and Edinburgh, he became a Fellow of Faculty in 1920. After naval service in the First World War he specialised in obstetrics and gynaecology and cultivated an interest in tropical diseases. He served on the Honorary staff of the Samaritan and Hahneman Hospitals in Liverpool, then developed a large obstetric consulting practice in Llandudno. See *Faculty Minutes*, 4 March 1957, RCPSG, 1/1/4/16, for obituary.
[32] Letter to Faculty from Dr S. Saxon Barton, 31 January 1939, *Council Minutes* 17 February 1939, RCPSG, 1/1/3/10. (My emphasis.)
[33] See Chapter 1, above.
[34] Letter from Saxon Barton to Faculty, 3 February 1939, *Council Minutes*, 17 February 1939.

of the Border'.[35] The rebuilding of the reputation of the Fellowship must begin at home: it must be made to 'matter in Glasgow'.[36]

Relaunching the Fellowship

Moves were already afoot, initiated by prescient Fellows, to remedy this potentially disastrous situation. In February 1939 the Faculty successfully applied to the GMC to change the registered abbreviation for the Fellowship. Up till now the abbreviation was FRFPS Glas., but this gave no indication of whether the Fellow had qualified in respect of Medicine or Surgery. In an era of increasing specialisation this made it difficult for the Fellowship to be defended as a higher qualification for senior hospital appointments, as it was not immediately apparent in which department of Medicine the holder's expertise lay. The President (John Souttar McKendrick) wrote to the GMC Registrar in February 1939 asking for permission to include an indication of department of Medicine in the Fellowship abbreviations on the Medical Register, as, 'it is felt that the addition suggested is desirable both as an indication to the public and to the Electing Bodies for hospital and other appointments'.[37] In a letter of 14 March, the GMC approved the change to the new abbreviations of FRFPS Glasg. (P) and FRFPS Glasg. (S),[38] which, though unwieldy, at least made it possible to argue that the Fellowship should be a recognised qualification for honorary hospital appointments as physician or surgeon.

As Saxton Barton had suggested, the next crucial step was to reinstate the Fellowship as the prerequisite qualification for senior hospital appointments in Glasgow. This challenge was taken up by the new President Roy Frew Young in early 1941.[39] A committee of Council was asked to draw up a memorandum to be sent to the chairmen and secretaries of all the major Glasgow hospitals.[40] The following version was dispatched after approval at

[35] Ibid.

[36] Ibid.

[37] Letter from PRFPSG (McKendrick) to GMC Registrar (Michael Heseltine), 20 February 1939, insert with *Council Minutes*, 17 February 1939, RCPSG, 1/1/3/10.

[38] Letter from Heseltine to Secretary of the RFPSG, 14 March 1939, with *Council Minutes*, 16 March 1939.

[39] Roy Frew Young (1879–1948). BA and MB, BChir Cambridge and Glasgow 1909. Pursued a surgical career at the GWI, rising to Visiting Surgeon, and then, on his retirement in 1944, to Honorary Consulting Surgeon. President 1940–42.

[40] The memorandum was sent to the Elder Cottage Hospital, the Glasgow Eye Infirmary, the Glasgow Hospital for Diseases of the Ear, Nose and Throat, the Lock Hospital, the Royal Mental Hospital, GRI, the Royal Maternity and Women's Hospital, the Royal Hospital for Sick Children, the Royal Samaritan Hospital for Women, the Victoria Infirmary and the GWI.

the Faculty meeting of 5 May 1941. Its language demonstrates an awareness that the Fellowship must be made relevant to the coming national and regional reorganisation of hospital work, and directly argued for the examination to be adopted as a useful staging-post on the emerging training ladder to consultant status:

> At the present time when there is taking place considerable discussion regarding the future of hospital practice the Council of the Royal Faculty of Physicians and Surgeons of Glasgow has been considering the relations of the Faculty with the hospitals in the West of Scotland. In England and elsewhere it is an established rule that any candidate for a senior appointment to the medical or surgical staff of any of the teaching or other large hospitals must hold the Higher Qualification of one of the Royal Colleges of Physicians and Surgeons. In Glasgow prior to 1914 few senior appointments in our teaching infirmaries were held except by those who were Fellows of the Royal Faculty … which for nearly three hundred and fifty years has been closely associated with the development of the voluntary hospitals. Since the last war, this policy, for various reasons, has been allowed to lapse. It is of obvious advantage to our infirmaries that the senior medical and surgical posts should be held by men who, in taking a Higher Qualification, have proved both their interest and ability in medicine and surgery. This would also be of great benefit to the younger men in encouraging them to reach a higher standard of work in preparing for the Fellowship Examination. Such a procedure then would be of material benefit to our infirmaries and their medical staffs. In view of these considerations it is suggested that for the posts of assistant physician and assistant surgeon or for equivalent appointments special attention should be paid by Selection Boards to the possession by the candidates of the Fellowship of the Royal Faculty of Physicians and Surgeons of Glasgow.[41]

When Young left the office of President, at the annual general meeting in November 1942, he was able to report that this memorandum had achieved its purpose. In his presidential report on the past year he stated that: 'We have now had assurances from practically all the hospitals that the possession of the Glasgow Fellowship in Medicine or Surgery, will be expected of all those who hold the more senior appointments'.[42]

The original draft of the memorandum to the hospitals had argued that the Fellowship was particularly appropriate as a prerequisite to senior appointments in Glasgow hospitals because it was 'a searching test of clinical ability, with which the infirmary is primarily concerned'. As such it compared favourably with a higher university degree, 'which indicates a certain ability

It was also sent to the MOH for Glasgow (Alexander MacGregor), and the Faculty's representatives on the boards of governors of the above hospitals.

[41] *Faculty Minutes*, 5 May 1941, RCPSG, 1/1/1/16.
[42] Ibid., 2 November 1942, RCPSG, 1/1/1/17.

to carry out original work'. This was 'complementary' to, but 'not a substitute for' the Fellowship.[43] These discarded comments tell us a lot about the ideology of medicine which the Faculty Council shared and saw the Fellowship as part of. It was an ideology of the primacy of clinical experience. Even in 1939 senior appointments to Glasgow hospitals (including teaching appointments) were evidently selected on the basis of clinical acumen, rather than the ability to carry out 'original work'. Clinical experience was paramount and scientific training and the ability to conduct scientific clinical research in the wards was undervalued. Glasgow was perhaps peculiar in that such an old-fashioned medical ideology had survived as long as the late 1930s.

That these phrases were taken out of the final memorandum to the hospitals indicates, however, that the Faculty was not only aware of how the coming Health Service would change the role of the Fellowship, but also that the Faculty was sensitive not to be perceived as publicly setting its face against modern, scientific, specialised, academic medicine. It would have been foolhardy to do so given the nature of the recent academic clinical appointments made by the University, which were part of a deliberate policy of bringing the new medicine into the hospitals. Changes made to the Fellowship regulations during 1942 also reflected this sensitivity to changes in medicine by introducing a much broader specialist element to the examination structure, which had been unsuccessfully attempted in 1924.[44] In 1942 a new section of the regulations was introduced to deal with 'those candidates professing a specialty'.[45] The two-part examination for such candidates still required that they take a written, a clinical and an oral examination in General Medicine or Surgery, but now allowed them, in the first part to specialise:

> A Candidate for the Fellowship who presents satisfactory evidence of having practised a specialty for at least seven years may submit himself for examination in that special subject if it be approved by the Council of the Faculty. The examination shall consist of (1) a written, a clinical and an oral examination in the special subject (the examination may be modified in accordance with the requirements of the subject).[46]

[43] Original draft of memorandum is in ibid., 3 March 1941.

[44] *Regulations of the Royal Faculty of Physicians and Surgeons of Glasgow*, bound, annotated version, RCPSG, 1/5/1/12. Compare this with the final version RCPSG, 1/5/1/11, in which the paragraph on specialist entry has been replaced by one continuing the privileges for service in the First World War. The 1924 draft regulations thus show that reform was blocked in that year and that the post-1890 format for the Fellowship examination was retained intact.

[45] 'RFPSG: Resolutions Proposed by the Council to Amend the Regulations Regarding the Ordinary Fellowship', insert with *Faculty Minutes*, 7 September 1942, RCPSG, 1/1/1/16. The Council's resolutions were unanimously approved at this meeting.

[46] Ibid.

While it had been possible for candidates to take one part of the examination in one from a list of specified subjects (for instance Ophthalmology, Aural, Laryngeal and Nasal Surgery, or Dermatology) since the 1890 reform of the Fellowship, this was the first time that the specialist component was widened to include other specialties.

Further moves were also made during Young's presidency to improve the public image of the Faculty. A series of postgraduate lectures was initiated by a Faculty committee consisting of J. Scouler Buchanan, Noah Morris and Stanley Graham.[47] The lectures were arranged by pooling the resources of the Burns Fund, the Finlayson Memorial Lectureship and the Watson Prize Fund. Commenting on their success in his 1942 presidential report, Young noted that it 'seems to me a pointer to the part which the Royal Faculty should play in future in Post-Graduate work in the city'.[48] Young delivered the opening lecture, which was largely an advertisement for the historical place of the Faculty in the medical life of Glasgow, and arranged to have it published, for further publicity, in the GMJ for December 1941.[49] In it he took the opportunity publicly to assert to the medical readership that 'The Fellowship itself, in Medicine and Surgery, is now rightly recognised as the higher degree which is regarded almost as an essential in any one seeking the higher posts in our hospitals'.[50] There were also talks on wartime cardiology, facial injuries in war, haemorrhagic states, bronchiectasis, prematurity, the management of closed head injuries and wartime abdominal injuries.[51]

This lecture series became a regular annual fixture. It was the first move into providing postgraduate teaching for the Fellowship by the Faculty, and it thus reinforced the campaign to relaunch the Fellowship with a practical contribution to the training dilemma of the intending consultant. Along with

[47] Scouler Buchanan was one of the last great general surgeons. He graduated MB, ChB from Glasgow in 1912 and he held a succession of surgical appointments at the GWI rising to Visiting Surgeon in 1936. In 1938 he became Honorary Lecturer in Surgery at the University of Glasgow. He served for several years on the RFPSG Council, becoming Visitor in 1946. Morris was one of the new breed of scientist-clinicians. He gained a Glasgow science degree in 1913 (with special distinction in physiology) and graduated MB, ChB in 1915. He was assistant to Noel Paton in the University Physiology Department, had a general practice and became Professor of Physiology at Anderson's. In 1921 he became Lecturer in Chemical Pathology and Biochemist at the Royal Hospital for Sick Children. He succeeded Stockman as Regius Professor of Materia Medica in 1937.

[48] *Faculty Minutes*, 2 November 1942.

[49] Roy F. Young, 'The Royal Faculty of Physicians and Surgeons and its Contributions to Medicine', GMJ, 7th Series, 18 (1941), pp. 159–68.

[50] Ibid., p. 168.

[51] Ibid., p. 122.

the meetings of the Royal Medico-Chirurgical Society of Glasgow (RMCSG), which provided both lectures, debates and case nights (clinical demonstrations, for instance in Surgery, at the GWI) [52] – the lecture series provided a forum for doctors from different specialties to keep up with recent developments.

In spite of these positive developments, the Faculty's financial position was still growing steadily worse. The problem was not helped by the reduction in candidates caused by so many medical students being away on active service, a difficulty which the Faculty had also experienced during the First World War. In the financial year ending 30 September 1941 the Faculty had incurred a massive deficit of £1131 19s. 11d. This was an increase even on the previous year's figure of just over £888.[53] The income from the TQ had been particularly badly hit by the war (down over £175 from the previous year, and about £400 down on the usual pre-war figure). Only the number of candidates for the Dental Licence remained steady at forty-eight. What had crippled the Faculty, however, was the year's exceptional wartime expenditure: £500 on carrying out the terms of the Fire Prevention Order and a further £353 on special insurance. Expenditure had thus been reduced on all possible areas of Faculty activity, but the Honorary Treasurer still had to deliver a gloomy prognosis:

> Leaving out of account minor economies there are two courses open to the Faculty. One that matters should be left to drift in view of the fact that War conditions will not last for ever and that when the hostilities cease the financial affairs of the Faculty will in all probability right themselves. The Second course is to face the problem now and balance our income and expenditure irrespective of what curtailment of the Faculty's activities would be involved.[54]

In December the Faculty was forced to ask Fellows for donations to ease the deficit or for gifts of journals to the library to save purchases. The situation was marginally improved by the end of 1942, when the deficit stood at just over £673. The Treasurer pointed out that the financial crisis forced Fellows to fundamentally reexamine the purpose of the Faculty as a medical corporation:

> any further reduction of expenditure must be drastic to be effective. Even so the pruning that might have to be done to balance the accounts might be so drastic as to endanger the very existence of the Faculty. We must always bear in mind

[52] *Minutes of the RMCSG, 1937–48*, RCPSG, 3/1/4.

[53] *Faculty Minutes*, 2 December 1940, RCPSG, 1/1/1/16.

[54] Financial details (1941) and quotation from *Faculty Minutes*, 3 November 1941 AGM, RCPSG, 1/1/1/16.

that the Faculty is a medical corporation and in the eyes of the public reflects the profession. Accordingly a certain dignity is, to my mind, essential and particularly so when the Faculty is so precariously situated financially. The matter cannot be approached purely from a business point of view. The great difficulty of the Finance Committee is to reconcile the business aspects with what I shall call, for want of a better word, [the] ethical aspect.[55]

This reexamination was already in progress, but Faculty minds were focused by the precarious financial position and also by important interlinked local and national developments in the following five years. Hector Hetherington was in the middle of a reorganisation of Glasgow medicine which would make the University the unitary centre of undergraduate medical education, bring an academic agenda of teaching and research to hospital practice, and offer the Faculty a new professional role as the local postgraduate teaching and examining centre to supply highly trained staff for the hospitals. Hetherington's plans both foreshadowed and were underlined by government reconstruction plans which had begun to focus on the provision of a National Health Service organised around specialist consultants, and which were soon to include full reviews of policy on hospital organisation and medical education. The resulting initiatives were to institutionalise patterns of organisational and educational structures which had already been evolving for many years within the wider medical community. These were to have a dramatic effect on the role of the Faculty, taking it much further towards being a postgraduate examining body, a new direction that it had already begun to move in.

Hetherington and the Modernisation of Glasgow Medicine

In Glasgow, Hetherington and his allies (notably the MOH Alexander Mac-Gregor, who became the first Chairman of the Western Regional Hospital Board in 1948, and George Wishart) set about reestablishing the University Medical School as a centre of progressive teaching and research.[56] This involved viewing the University as the natural educational centre of the region and managing educational resources on a regional basis. In medicine this involved strengthening the formal links between the university and the existing voluntary teaching hospitals (GRI, GWI and VI), and establishing new links with the upgraded municipal general hospitals (Stobhill and the Southern General). By a series of key appointments Hetherington introduced

---

[55] *Faculty Minutes,* 2 November 1942 AGM, Honorary Treasurer's Report, RCPSG, 1/1/1/17.
[56] For more on this see Andrew Hull, 'Hector's House: Sir Hector Hetherington and the Academicisation of Glasgow Medicine before the NHS' (*Medical History,* forthcoming).

the university ethos of linked teaching and research into the hospitals. Full-time professorial clinical units with control of beds were established in the Western Infirmary (where J. W. McNee was appointed Professor of Medicine in 1937 and for whom adjacent laboratory facilities were provided at a new Institute of Medicine in 1938; and also where Charles Illingworth was appointed Professor of Surgery in 1939), at Stobhill Municipal Hospital (where the biochemist-clinician Noah Morris took up the Chair of Materia Medica in 1937, and for whom a new biochemical laboratory was built on the site), and at the Royal Infirmary (where L. J. Davis became Professor of Medicine in 1945). These new academic units with full-time Professors followed the Flexnerian model: programmes of clinical research were initiated and young researchers were attracted in to form a research school, and clinical teaching was also carried out.[57] With the exception of McNee, all of these men played a central role in the establishment of postgraduate education for consultant specialists in Glasgow, largely through their activities in the Royal Faculty, which, during Hetherington's time as Principal and Vice-Chancellor of the University came to function as the university's postgraduate teaching and examining arm. In the context of wartime plans for a National Health Service based on a regional plan which confirmed the local importance of the university in medical education and research, Hetherington (and his newly appointed professorial allies) attempted to further extend university influence over the hospitals by taking control of the payment of all clinical teachers in the hospitals, and by seeking to ensure strong university representation on the proposed Medical Education and Appointments committees of the new regional board. There were also university instigated attempts to informally seed the peripheral hospitals (for instance the Southern General) with consultants from the main teaching hospitals to prepare them for use as undergraduate and postgraduate teaching centres, and to begin programmes of clinical research there, in advance of the Appointed Day.[58]

This new regime in Glasgow Medicine needed organised postgraduate training to supply highly-trained staff to the consultant-based hospitals. The University was to have its hands full with undergraduate teaching and examining. The Faculty, in which Hetherington's new men occupied influential

---

[57] On this see, for example John Pickstone, Roger Cooter and Caroline Murphy, 'Exploring "Clinical Research": Academic Medicine and the Clinicians in Early Twentieth-Century Britain', paper presented at conference on Science in Modern Medicine, Manchester, 19–21 April 1985; and Ministry of Health, *Report of the Inter-Departmental Committee on Medical Schools* (London, 1944) (The Goodenough Report), especially chapter 4, 'Staffing', section on 'Teaching Staff in Clinical Departments, Including Pathology', pp. 79–91.

[58] Interview with Dr Gavin Shaw by Andrew Hull, 29 October 1997.

positions, and which was to be left with no professional role after the planned closure of the extra-mural schools, was the natural focus for postgraduate education.

The Faculty and the National Health Service

The imminent national reorganisation of medical education and practice was to follow the pattern already established in Glasgow by Hetherington. Institutional arrangements laid stress on the University as the regional centre of medical education, and hospital practice was to be modernised by being permeated with an academic ethos via clinical units.[59] The Faculty had a cautious attitude towards these national plans. While the key players in Faculty and University were increasingly the same people, the Faculty had a history of representing the local clinical elite and of resisting University encroachments into the hospitals. The reforms which led to the NHS embodied a direct challenge to the existing social and intellectual authority of the existing clinical elite, threatening to replace the hospitals' focus on clinical work and the hegemony of a part-time clinical elite, with a new focus on teaching and research as the basis for a new academic hospital elite. This contrast of viewpoints was clearly brought out in an address given by Sir Alfred Webb-Johnson to the Faculty on the occasion of admission to the Honorary Fellowship in 1945. He identified the practical, clinical element in medicine with the medical corporations and saw this as under threat in the integrated package of education and organisation of the new health service from the academic, scientific element. An elite London surgeon, and President of the Royal College of Surgeons from 1941–48, he strongly defended the 'incommunicable' clinical art as the essence of medicine (and, as Christopher Lawrence has argued,[60] as the basis of his own social position):

> Those who continually emphasize the need of scientific and theoretical training are inclined to neglect the practical and vocational side and to forget that the practice of medicine is largely an art. The practice of medicine is essentially vocational and not academic and a large proportion of the most useful and successful practitioners in the country have never obtained, and do not desire to obtain, a University degree.

He was moved to continue on this theme with increasing hyperbole:

> Now that the prestige of science is so high the statement that a great part of medicine still retains the status of an art is often made with a note of apology.

[59] See Goodenough Report.
[60] Lawrence, 'Incommunicable Knowledge'.

Nothing could be less justified by a realistic sense of cultural values. The method of the practical art was the first instrument forged by man for the subjugation of chaos. At the dawn of civilization the preservation of knowledge was far more important than its discovery. Methods of prevision must be very strictly subject to the art of medicine if they are not to become a real snare. The truth appears to be that what the user of a practical art needs is less the strict and limited instrument of scientific method than what may be called a soundly cultivated judgement. The ancient and honourable art of medicine is being increasingly and inevitably pressed on by applied science ... It remains, however, the backbone of medical practice and indispensable to mankind. There is therefore an especial need today that its *characteristic mode of activity* should be understood, and should not be confused with those of the other elements that make up the complex of medicine.[61]

The address was published in the *GMJ* with Faculty support.

Like the Faculty itself, Webb-Johnson was not against science in medicine,[62] nor was he against a National Health Service,[63] but he did not want the leadership of the profession to pass from the elite clinicians of the Royal Colleges to academic clinicians committed to clinical research under the new arrangements. However, as we have already noted (and as the institution's key officers had perceived in the run up to 1858 and at all moments of change before and since), the Faculty was also keenly aware that change was imminent and of the necessity of cannily repositioning itself and its professional role to become a vital player in the new situation.

The Faculty's official pronouncements on the plans for the new Health Service reflected this ambivalent stance. In November 1942 the Faculty submitted its evidence to the Goodenough Committee in the form of a memorandum on 'The Future of Medical Teaching in Glasgow', written by J. H. Wright.[64] As he put it, this outlined the Faculty's views on 'just how

[61] Address by Sir Alfred Webb-Johnson to the RFPSG, *Faculty Minutes*, 4 June 1945, RCPSG, 1/1/1/17. Also printed in the *GMJ*, seventh series, 26, August 1945 as 'The Royal Medical Corporations', pp. 33–38.

[62] On the contrary, he was instrumental in obtaining funds for the endowment of professorships in the basic sciences as Dean of the Middlesex Hospital Medical School. See Z. Cope, *The Royal College of Surgeons of England: A History* (London, 1959), p. 207.

[63] Along with Lord Moran, President of the Royal College of Physicians, he was one of Bevan's elite consultant allies who came out in public support of the NHS (as long as certain guarantees were given to doctors), and in favour of the BMA negotiating with the Minister of Health, during the most difficult period of 1947. Bevan was seen by many in the BMA as using such allies to divide professional opposition to the NHS. He was also believed to have bought them off with the promise of private beds in NHS hospitals. See Foot, *Bevan*, chapters 1–4, especially, pp. 162–65.

[64] Interview with Dr Joseph Wright, 8 February 1986 by Dr Peter MacKenzie, RCPSG, 18/20,

hospitals should be run'.[65] The memorandum stated that the Faculty approved of the linking of voluntary and municipal hospitals for medical education, but believed that the best way of accomplishing this was, 'individual autonomy within a coordinated regional scheme'.[66] Junior but not senior appointments should be full-time, and under the control of the regional board, so that the traditional working patterns and independence of the clinical elite should be protected. The Faculty stressed that its higher qualifications of the royal medical corporations should be used in the new system to distinguish consultants: 'In the selection of Chiefs and senior staff a qualification such as the Fellowship of the Royal Faculty of Physicians and Surgeons, or the Fellowship of the Royal Colleges of Physicians and Surgeons should be considered essential'.[67] But it was in reply to the question of whether or not the University degree should become the required registrable qualification that the Faculty memorandum reserved its fullest response. Such a move would not be 'in the public interest'. The Faculty reminded the interdepartmental committee of the long duration of its involvement in undergraduate licensing, stating that: 'In the opinion of the Royal Faculty the creation of a University monopoly would be a retrograde step'. Not only would the abolition of the extra-mural schools and the institution of a University monopoly deprive students of choice, and possibly discriminate

transcript, p. 35. Joseph Houston Wright (1899–1989) graduated MB, ChB from Glasgow in 1922 and became house physician to Dr W. K. Hunter at the GRI with a general practice in Bath Street. He became assistant physician to Dr A. W. Harrington at the Royal in 1927. He also became the hospital's cardiographer in 1929, the beginning of a lifetime's interest in cardiology. He became a Fellow in 1932 and continued with Harrington, becoming his First Assistant (with full ward access) when Harrington got the Muirhead Chair of Medicine. When L. J. Davis succeeded him in 1945, Wright and his immediate colleagues moved to another unit to give the new professor a chance to build his own distinctive department. Wright, who by now had a large consulting practice, worked in a small investigative ward on hypertension, but in 1948 he was promoted to full charge of a general medical unit in wards 6 and 7. Here he built up an academic department in all but name, having accepted the new thinking on the role of senior clinical staff that had come to the GRI with Davis's appointment. The unit began cardiac catheterisation and did mostly cardiological work. Wright encouraged clinical research and built up a team of keen young researchers around him. He was Senior Consultant Physician and also Clinical Lecturer to the University. He was one of the original members of the WRHB, and in 1964 was the author of the influential Wright Report. See Robert Fife, 'Twentieth-Century Scottish Physicians: Dr J. H. Wright', copy of article gifted by the author.

[65] Ibid.

[66] 'Royal Faculty of Physicians and Surgeons of Glasgow: Memorandum of Evidence on Medical Schools', 26 October 1942, insert with *Council Minutes*, 19 October 1942, RCPSG, 1/1/3/10.

[67] Ibid.

against Empire students, it would have fatal consequences for the medical corporations:

> The qualifications granted by the Royal Faculty in cooperation with the Royal College of Surgeons of Edinburgh have not been attacked and the Royal Faculty fails to appreciate why it should be suggested that non-university qualifications should cease to be registrable. To deny the Medical Corporations the right to grant registrable qualifications, as they are at present by Status entitled to do so, would lower the status of these corporations and would endanger their existence. The Final Examination for the Scottish Triple Qualification is of the same standard as the corresponding examination of the University of Glasgow, and the clinical examiners are frequently the same persons.

The memorandum also emphasised that the corporations and their examinations stood for the practical clinical element in medicine, still seen in many high London medical circles as the touchstone of medical education and practice, as Webb-Johnson's above-quoted address demonstrates:

> The Medical Faculty of a University includes relatively few members who are actually engaged in the clinical practice of their profession. The Medical Corporations, on the other hand, are representative of the best and widest cross section of the medical profession, and it is eminently desirable that entrance to the medical profession should be, in common with other professions, largely in the hands of those who have attained responsible positions in the practice of the profession.[68]

This attitude of fear and suspicion of plans for the universities was not shared by the Edinburgh Physicians. Their approach was captured by the comments of a senior Fellow who commented on the proposed closure of the extra-mural schools in the Goodenough Report, 'that neither sentimental nor historical considerations' should be overvalued, and that 'harmony with the University in the general interests of medical education' should be maintained.[69]

The Faculty memorandum ended with more of an eye to the future, arguing that the *higher* qualifications of the medical corporations (again

---

[68] Ibid.

[69] Craig, *History RCPEd*, p. 721, citing Minutes of the RCPEd, 22 April 1943. They seemed, however, to have later come round (with the Edinburgh Surgeons) to the Faculty's point of view. All three bodies supported a joint memorandum in April 1945 opposing the closure of the extra-murals in Scotland on the basis that they provided a cheaper alternative for students, were not (unlike the universities) reluctant to take overseas students, and were necessary because the universities could not teach all the current crop of medical students. See 'Draft Memorandum by the Royal Scottish Medical Corporations on the Report of the Inter-Departmental Committee on Medical Schools', April 1945, RCPSG, 1/5/18.

these were described as, 'chiefly concerned with the practical aspects of medicine and surgery') should be used to answer the difficult question of 'who is a consultant?'.[70] This latter theme of how should consultants be so designated was addressed in a joint memorandum of the three Scottish Royal Medical Corporations and the Royal College of Obstetricians and Gynaecologists, in July 1943.[71] This document shows remarkable prescience, or is rather, perhaps, a testament to the enduring, but slow-acting, influence of the corporations in this self-governing profession of medicine. Postgraduate education and training have now (after an intervening period of doubt and debate) been formally organised along lines recommended in this joint memorandum. The document shows just how clearly the corporations had understood the gravity of the changes facing them as early as 1943, and that they had then already mapped out a strategy for ensuring their survival.

The joint memorandum argued that whatever the organisational basis of the new health service, specialist and consultant services would be 'an integral and important part'.[72] In order that the service should be efficient and competent, 'it is clear that some machinery must be set up to regulate the training of specialists',[73] since, 'Proposed changes in the conditions of medical practice call for further organization of the training, examination and designation of clinical specialists in medicine, surgery and midwifery, and of their subdivisions'.[74] Rather than, 'be faced with the setting up of a great number of new diplomas each awarded by a specially chartered or incorporated body with a limited field of activity', or entrusting the job of laying down courses, setting up examining boards and awarding diplomas to the GMC, the committee, unsurprisingly, thought that,

> existing licensing bodies, and in particular the Royal Medical Corporations of England, Scotland and Ireland [should] come to the assistance of the Government and accept responsibility of prescribing the proper training for the specialist, of examining him, and of appropriately designating him when his training is complete and he has passed an examination in a satisfactory manner.[75]

[70] Ibid.

[71] 'Memorandum of the Joint Committee of Representatives of the Three Royal Scottish Medical Corporations and of the Royal College of Obstetricians and Gynaecologists on Specialist Training and Status', 16 July 1943, RCPSG, 1/5/19.

[72] The memorialists chose to use the term specialist throughout the memorandum, preferring the term to consultant, though they meant both, while being aware of the differences between them.

[73] Ibid., p. 1.

[74] Ibid., p. 2

[75] Ibid.

It was suggested that the training of the specialist should consist of two parts. First would come clinical experience in resident hospital appointments, and study of the basic medical sciences. This part would end with an examination for a higher qualification of one of the royal medical corporations. The second stage would consist of a further period of hospital experience in a higher grade hospital post, at the end of which the specialist or consultant diploma would be conferred which specified in which field the holder had been trained. After getting such a specialist diploma, the individual would then be eligible for permanent posts on the staff of approved hospitals.[76] The memorandum recommended that the GMC should take on the role of a 'controlling authority with powers of recommendation and supervision' to manage postgraduate education in the same way that it had now long managed undergraduate education. It should also fall to the GMC to compile a list of bona fide specialists (similar to the Medical Register) if it was decided that such a list was necessary.[77]

It was not until 1968 when a consensus formed within the profession around the recommendations of the Todd Report that postgraduate training was so divided into early general (ending with a higher qualification), and then later specialist periods. The GMC did not take over responsibility for managing postgraduate education in the same way as undergraduate education until 1978, and it was not until 1997 (with the promulgation of the Specialist Medical Order on 1 January) that accreditation of specialists and a register became mandatory. The memorandum is a mark, nevertheless, of how early the corporations' responses to the challenge of the medical reforms of the 1940s and 1950s were very far advanced, showing a sophisticated grasp of the concomitant change in the role of the corporations to ostensibly postgraduate teaching and examining bodies.

This kind of foresight should be set against, and is more important for our concerns here than, the Faculty's opposition to the Health Service as an interlinked social and epistemological threat to the established clinical elite. This was part of a common medical response to the introduction of the NHS by both GPs and consultants who all feared the consequences of a state (or even worse, a local authority) service, which has been well-documented elsewhere.[78] This opposition, outlined above, resurfaced periodically throughout the national debate leading up to the Appointed Day. For instance a Faculty memorandum of July 1944 suggested that the Health Service should be restricted to those already covered by the NHI scheme and their depend-

[76] Ibid., pp. 2–3.
[77] Ibid., p. 3.
[78] See the references in n. 4 above.

ants, and that, even then, doctors should not be forced to be full-time state employees. It argued that present plans,

> will inevitably result in the elimination of two of the best features of British medicine, independent general practice and the Voluntary Hospitals, and the establishment of a State controlled medical service in their stead. Admittedly, the present arrangements could be improved by coordination and some financial assistance from the State, but these do not necessitate bureaucratic control which is contrary to the tradition and genius of the people of this country. Under the new proposals the doctor will become a State or local authority official. This will destroy the personal relationship between doctor and patient.[79]

First Steps in Postgraduate Education

By 1944 the Faculty, as well as planning for the future of postgraduate education under the NHS, had begun to take practical steps to become part of this future. Apart from the postgraduate lecture series begun in 1942, there was still almost no organised provision for those studying for higher qualifications. A contemporary candidate for the Glasgow Fellowship recently remembered the paucity of organised postgraduate training in the early 1940s:

> I graduated in 1943 and after several house jobs at the Southern General Hospital I was appointed to the post of resident Medical Registrar at that hospital. My 'chief', Dr Snodgrass advised me to sit the Glasgow Fellowship (FRFPSG) which I obtained in 1946. At that time there were no post-graduate courses of any kind – one had to devise one's own course of study, which involved subscribing to the *Lancet* and *BMJ* – plus purchase of textbooks. I also had access to the *Quarterly Journal of Medicine* and to *JAMA*. In addition one 'picked other people's brains'! I do not remember any structured courses at the College [Faculty], though there were occasional evening lectures ... open to any interested doctors. Study leave was quite unheard of. As you can imagine, it was very difficult to study while living on call every second day and night and every other weekend. I thought it a great advance when post-graduate courses were instigated.[80]

---

[79] 'Observations on the Government's Proposals for a National Health Service', July 1944, RCPSG, 1/5/17. The document is signed by the President (James H. MacDonald) and Visitor (William A. Sewell).

[80] Letter from Dr A. M. K. Barron to Andrew Hull, 14 February 1997. I am grateful to Dr Barron for permission to reproduce her remarks here. After Dr Barron had come to the end of her contract as Registrar at the Southern General in 1947 (by which time she had gained her Glasgow Fellowship), she went to Snodgrass and asked his advice. She was keen to become a consultant physician with a special interest in cardiology, after she had completed a clinical research project on the heart in pregnancy. This was the period (1943–47) when the Southern General was being turned from a Poor Law institution into a teaching hospital. Snodgrass replied that she should forget all about such ambitions as there would not be female consultants

In January 1944 the Department of Health for Scotland had contacted the Scottish Universities regarding the imminent return of large numbers of demobilised medical officers who had been away on war service. These returning practitioners would need intensive refresher courses before returning to their civilian medical work. The scale of the problem was large. The department estimated that there would be between five and six hundred prewar general practitioners who might need refresher courses before returning to practice, but the more urgent problem was the arrangements for the nine hundred or a thousand younger graduates who had gone into service after holding junior resident hospital appointments for only six months and who had no previous experience of general practice.[81] The department asked the universities to appoint officers to be deans of postgraduate medical teaching and convenors of new University Medical Faculty committees charged with the organisation of postgraduate courses. UGC funds were to be channelled to the universities to defray the cost of these courses,[82] and the universities would have independent control of these funds and be able to subsidise the activities of any other bodies they might choose to involve. In Glasgow the medical faculty decided to appoint Professor James Hendry[83] as part-time Postgraduate Dean and Convenor of the Postgraduate Committee and initially to restrict membership of the committee to Medical Faculty members on the Board of the existing Glasgow Postgraduate Medical Association.[84] This meant Hendry,

in this part of the world for a considerable time! He advised her to marry instead, which she duly did to a GP (after some GP work herself). However, she returned to hospital practice in 1961 at the Greenock Eye Infirmary, where she did attain consultant status in 1973 as an ophthalmic surgeon on the basis of her experience and Fellowship. Details from a telephone conversation with Dr Barron by Andrew Hull, 1 July 1998.

[81] *Minute Book of the Glasgow Postgraduate Medical Association*, meeting of 28 March 1944, GUABRC, Med 2/1.

[82] *Faculty Minutes*, President's Report for 1946, 4 November 1946, RCPSG, 1/1/1/17.

[83] James Hendry (1886–1945). After graduating in medicine and science from Glasgow in 1913, Hendry took up the post of Second Assistant to Dr Munro Kerr, since 1911 the Muirhead Professor of Obstetrics at the Royal. He succeeded Munro Kerr in the Chair in 1927, and was also surgeon in charge of the Royal's gynaecological department. In 1943 he became Regius Professor of Midwifery and Medical Director of the Royal Maternity Hospital, while retaining his position as Gynaecological Surgeon to the Royal. As well as being a Fellow of the Glasgow Faculty, he was also an active Fellow of the RCOG, his membership of the Goodenough Committee led naturally to him being the first Postgraduate Director at Glasgow University, when the post was still part-time. Like Noah Morris, Hendry was part of a new breed of Glasgow practitioners, and, again like Morris, his early loss was felt deeply among the medical community at the time, and by future historians. Obituary, *GMJ*, 144 (1945), pp. 83–84.

[84] *Minutes of the Medical Faculty of the University of Glasgow*, 11 January 1944, GUABRC, Med 1/13.

Charles Fleming,[85] McNee and the Dean of the Medical Faculty, George Wishart. On 7 March Hendry reported that he and Wishart had attended a conference in Edinburgh on postgraduate courses for demobbed MOs at which it had been made clear that most of the local organisation would be handled by the respective postgraduate committees. In the light of this Hendry recommended to the University Medical Faculty that the Committee be enlarged. He reported that the Department of Health had confirmed that it would leave,

> complete freedom to each University Faculty in regard to collaboration with other interested bodies and the University Court approved of the cooption of persons outwith the Faculty on the understanding that the Committee would be part of the organization of the Faculty of Medicine.[86]

The committee was duly expanded to include six members of the Medical Faculty (Professors Hendry, who remained Convenor, Fleming, McNee, Morris, Illingworth, and Wishart as Dean); four members of the RFPSG, Dr A. D. Briggs, the Medical Superintendent of Stobhill, and James Carslaw, the Secretary of the existing GPGMA.[87] The Faculty, which had been informed by President MacDonald on 6 March that the University was about to invite them to join the new postgraduate committee,[88] appointed William Sewell, David Smith, Snodgrass and Roy Frew Young as its own representatives,[89] although, of course, Hendry, Fleming, Morris and Illingworth were all active Fellows. While the new committee was thus nominally a University committee, the voice and interests of the Faculty were strongly represented on it. The committee met for the first time on 20 April 1944 to discuss the 'basic principles' of the postgraduate scheme to be put into operation for demobbed doctors, and to appoint a subcommittee to draw up more detailed plans.[90]

---

[85] Charles Mann Fleming succeeded Wishart as Dean of the Medical Faculty and Postgraduate Dean in 1959. In 1960 he was also created Professor of Administrative Medicine in the University of Glasgow because Hetherington felt that the Dean should have professorial authority. He held all three posts until 1971, and was also, like Hendry and Wishart before him, Convenor of the Post-Graduate Medical Education Committee from 1959.

[86] *Minutes of the Medical Faculty of the University of Glasgow,* 7 March 1944.

[87] Ibid. Carslaw had worked under Gairdner and Clare Cameron, and Stockman. He got his own wards as Visiting Physician at the GWI in 1920, had interests in rheumatic diseases and particularly in disorders of the blood, liver and spleen. He retired in 1934, but continued to act as Secretary to the GPGMA (a post which he had taken up in 1919) until the body was wound up in favour of the new University/Faculty GPGMEC in 1944.

[88] *Faculty Minutes,* 6 March 1944, RCPSG, 1/1/1/17.

[89] *Minutes of the Medical Faculty of the University of Glasgow,* 25 April 1944.

[90] Ibid.

Now that the University Postgraduate Medical Education Committee (GPGMEC) was up and running, the old Glasgow Postgraduate Medical Association decided to wind itself up in December 1945. On Carslaw's suggestion, the remaining funds (£138 5s. 5d.) were given to the Royal Faculty to help run its postgraduate lecture series.[91] On Hendry's death on 9 September 1945, Morris succeeded as convenor.

As well as refresher courses for returning GPs, or those younger students whose education had been interrupted by war service, and who wanted to become GPs, the new committee was also to select individuals to become consultants and specialists. These individuals would be given six month clinical attachments in relevant Glasgow hospital units. Both the GP and consultant schemes were funded by the Department of Health. After the Appointed Day, the selection of trainee specialists and the clinical attachment training scheme continued to be administered by the GPGMEC in Glasgow, with liaison with the new Western Regional Health Board and the individual boards of management of the hospitals. In other Scottish NHS regions this function was then fulfilled by the new Medical Education Committees (MEC) of the Regional Health Board (RHB).[92] In Glasgow the GPGMEC (and its predecessor) had a long history of cooperation between the University, the Faculty and the clinical teachers in the hospitals. Its unique status was underlined and strengthened under the NHS arrangements in Glasgow, where the MEC of the RHB confined itself to policy for undergraduate instruction in the hospitals. The importance of the GPGMEC was further enhanced in 1947 when Wishart was appointed the first full-time Director or Dean of Postgraduate Medical Education, as well as Dean of the Medical Faculty. Morris resigned as Convenor of the GPGMEC, and Wishart took over, so that there should be a single key individual in control of medical education in Glasgow.[93]

In June 1946 the Faculty, after consulting with the University, set aside two lecture rooms for the use of thirty graduate students, and in November the Burns Room was denoted a study-room for postgraduate candidates for the Fellowship and GPs on refresher courses, and, with the appointment of an extra librarian, the new library became a working library for postgraduate

[91] *Minute Book of the Glasgow Postgraduate Medical Association*, final meeting, 3 December 1945, GUABRC, Med 2/1; *Faculty Minutes*, President's Report for 1946, 4 November 1946, RCPSG, 1/1/1/17.

[92] See 'National Health Service: Regional Hospital Boards: Training of Specialists', RHB (S), insert with *Council Minutes*, 15 March 1949, RCPSG, 1/1/4/8.

[93] *Minutes of the Medical Faculty of the University of Glasgow*, 30 April, 29 October, 14 December 1946 and 29 April 1947, GUABRC, Med 1/4.

study. At this stage most of the postgraduates were GPs on retraining courses.[94] At the 1946 AGM the President (Geoffrey Balmanno Fleming) reported that, given the imminent demise of the extra-mural schools and the consequences for revenue from the TQ and LDS, Faculty involvement in postgraduate education was 'vital'. The Faculty now provided study rooms and the evening lecture series, and, underlining the necessary 'close co-operation' with the University on this matter, the new GPGMEC now met in Faculty Hall. Fleming concluded his final report as President emphasising the interlinked programme of postgraduate education and the revival of the Fellowship were the way out of the problems posed by the coming changes in the profession:

> We must continue to shoulder a considerable part of the burden of post-graduate teaching. I am convinced this is a development which can do nothing but good to the prestige and recognition of our Corporation and its Fellowship. It is most important that our Fellows should instil into the minds of their juniors in our hospitals that it is essential that they become Fellows if they wish to pull their weight in our Glasgow School.[95]

During 1947, however, Faculty officers became concerned about the level of expenditure on facilities for postgraduate education, in spite of the warnings of Walter Galbraith the Honorary Treasurer that such support must be given 'without stint' and reassurances that such expenditure would pay long-term dividends, and, in the short term, post-war increases in Fellowship income would offset the imminent disappearance of the TQ and LDS.[96] In April 1947 the Faculty communicated to the University that it, 'could not be expected to undertake such activities unless some financial aid was forthcoming'.[97] Fleming sent a report and estimates for the conversion of the classrooms and the employment of the extra librarian to the GPGMEC and the University Court,[98] and in May Faculty fears were eased when the Court announced it was prepared to give £1000 towards the Faculty's capital expenditure on transforming rooms for postgraduate use, and a further £1112 towards annual expenditure. The Council now sanctioned the refurbishments.[99]

The discussions which led to this successful outcome for the Faculty went

[94] *Faculty Minutes*, 18 June and 19 November 1946.

[95] This and preceding quotations in this paragraph are from 'President's Report', *Faculty Minutes*, 4 November 1946 AGM.

[96] 'Honorary Treasurer's Report', ibid.

[97] Ibid., 14 April 1947.

[98] *Council Minutes*, 18 March and 15 April 1947.

[99] *Faculty Minutes*, 20 May 1947.

on within the GPGMEC, but it was Hetherington's close ally George Wishart (doubtless with the authority of the Principal behind him) who argued the Faculty case directly with the University and finally secured the grant, which was to come from redirected University Grants Commission funds.[100] The interconnection of Faculty and University at this time is well illustrated by Wishart. He was at the same time Postgraduate and Medical Dean, Convenor of the GPGMEC, and the Principal's right-hand-man, *and* a member of the Faculty Council during 1947–48. The University continued to give the Faculty an annual £1000 grant for postgraduate education, which could be spent as the Faculty thought fit, right through the late 1940s and the 1950s. These funds were absolutely crucial to the Faculty's financial survival during the bedding-in period of postgraduate education and examination, when it could not be expected to bring financial return. They are another example of how Hetherington was keen to strike a bargain with the Faculty: if the Faculty's clinical elite allowed him to modernise medical practice, they would reap the reward of being supported by the University as its postgraduate teaching and examining arm, providing the highly-trained personnel which the new Glasgow Medical School needed.

The next year, 1948, saw the beginning of debate about the best means of organising postgraduate qualifications which was to concern the Faculty for at least the next twenty years. This was the problem contemporaneously described a 'multiple diplomatosis'.[101] This was the question of whether consultant specialist status should be indicated by the passing of separate specialist diplomas, or whether the higher qualifications of the royal medical corporations alone should indicate consultant specialist status. If the latter course was chosen the higher qualifications would need to include a number of specialist permutations. In the late 1940s there was no national policy on this question. The Faculty was keen that the Fellowship should be regarded as *the* passport to consultant and specialist status, and yet was also concerned that this qualification remain broadly-based in *general* medicine and surgery, in keeping with the role of the Faculty as the representative of the Glasgow clinical elite with its epistemological attachment to the primacy of general clinical experience. However, the situation was in a state of flux in the late 1940s and early 1950s (and did not resolve itself until after the Todd Report of 1968). Thus the Faculty had to be flexible and balance its attachment to general education with the moves of the other corporations.

---

[100] See interview with Joseph Wright, by Peter MacKenzie, 8 February 1985, RCPSG, 18/20.
[101] For this usage, see interviews with Professor Holmes Hutchison by Peter MacKenzie, 1985, RCPSG, 18/12, and with Professor Edward McGirr, RCPSG, 18/16.

In 1948 the English Colleges and the British Postgraduate Medical Federation (based around the Hammersmith Postgraduate School, but including
outlying English institutions) established a joint board to discuss and plan
all issues in postgraduate education and especially the granting of specialist
diplomas. By the early 1950s this conjoint board was offering diplomas in
Anaesthetics, Child Health, Industrial Health, Laryngology and Otology,
Medical Radio-Diagnosis, Medical Radio-Therapy, Ophthalmic Medicine
and Surgery, Physical Medicine, Public Health, Psychological Medicine, and
Tropical Medicine and Hygiene.[102] Hetherington commented on this move
that:

> It is obviously aimed at encouraging candidates to take London courses, thereby
> to encourage them to take the English College Diplomas, and so to gain the
> English Colleges an indirect control over entry into [the] specialist rank of the
> Hospital Service. There was no Scottish representation and its formation and
> activities cause considerable disquiet here in Scotland.[103]

Hetherington's attitude reflected the Goodenough Report: all new specialist
diplomas should be the responsibility of the royal medical corporations.
Responding to developments, the Faculty debated establishing specialist
diplomas in Industrial Health, Psychiatry, Radio-Diagnosis and Radio-
Therapy, and further discussed the practicality of diplomas in Anaesthetics,
Child Health, Ophthalmology and Oto-Rhino-Laryngology.[104] In all cases
there was considerable consultation both with the University and with the
Edinburgh Colleges, with the goal of establishing a joint diploma awarding
body to take on the London axis. However, the problems involved in such
moves were not only about putting together a united front against competition from London. These discussions continued into the early 1950s but
were ultimately unsuccessful.[105] A Faculty Council working paper from February 1948 argued that the need for specialist diplomas was created by the
intense competition among young graduates for the new salaried NHS consultant posts.[106] Most did not have time to take the training in general
medicine and surgery and in the basic sciences that was necessary before
taking a higher qualification. Hence there was a strong demand for an
alternative method of qualifying for consultant and specialist status. Since

[102] Hetherington, personal notes on 'Specialist Diplomas' (n. d. but early 1950s), Hetherington
Papers, GUABRC, DC8/1158.
[103] Ibid.
[104] *Faculty Minutes*, 1 March 1948.
[105] See *Minutes of the Medical Faculty of the University of Glasgow*, 6 March and 1 May 1951.
[106] 'Postgraduate Medical Diplomas', insert with *Council Minutes*, 27 February 1948, RCPSG,
1/1/4/7.

the problem of what constituted a consultant and specialist was not to be part of any new medical legislation, the English Colleges had attempted to control entry to higher hospital posts by establishing an official register of specialists, but had been stopped by the GMC. The 1950 Medical Act was supposed to have contained a clause giving this power to the GMC, but the clause was too controversial. For instance, the Faculty had objected to overall GMC control because having had undergraduate examination taken away by the universities, it did not want the GMC to take postgraduate examination just at the time that it was attempting to carve out a new professional role in this sphere.[107] This is the reason that GMC supervision of postgraduate education, though it seemed a natural and logical, not to mention sensible, extension of its role in undergraduate examination, was long delayed (until the Medical Act of 1978). This clause was dropped from the 1950 Medical Act to allow the Bill an easy passage into law.[108]

In the late 1940s and 1950s there was therefore a free and unregulated market (*laissez-faire* or *Lernfreiheit*) in postgraduate education, much as there had been in undergraduate education before the passing of the Medical Act of 1858. The corporations had thus to engage in the same kind of manoeuvrings that had spawned the Scottish DQs and the TQ in the nineteenth century. In contrast to the long and unspecified training necessary for a higher qualification from one of the royal medical corporations (which neither specified nor provided any structured training), the period of training for a specialist diploma was typically eighteen months after a well-structured course provided by the examining body. This, and the tendency of hospital authorities to ask only for such diplomas when advertising for consultant posts, had led to a dramatic rise in demand for them. As the 1948 Faculty working-paper stated, 'the combined effect of these factors militates in favour of a Diploma and against a Fellowship'.[109] Developing this theme it argued:

> The provision of courses of instruction and of examinations for Diplomas in special subjects is not without danger. The true object of the Diploma should be essentially preparatory and educative rather than vocational and the training should always include instruction in the basic sciences ... *Moreover it is, or should be, an essential preliminary for any one desirous of practising a speciality that he first undergo a period of training in general medicine or surgery.* The difficulty of reconciling such ideals with the insistent and widespread demand for early specialisation is obvious and there is a real danger that the Diploma will be used

[107] See *Council Minutes*, 2 January 1947, RCPSG, 1/1/4/6, Report of Council Committee on Draft Medical Bill.
[108] Hetherington, 'Specialist Diplomas'.
[109] 'Postgraduate Medical Diplomas', *Council Minutes*, 27 February 1948.

merely to enable the Diplomate to appear qualified to practise a specialty when in fact he has not been adequately trained.[110]

The Faculty instituted a Diploma in Industrial Health from June 1948,[111] but the Council rejected the idea of instituting a diploma in Anaesthetics. This latter was suggested by H. H. Pinkerton, the first Secretary and Treasurer of the Glasgow and West of Scotland Society of Anaesthetists. Pinkerton's request that Anaesthetics be included as one of the special subjects for the Fellowship examination had also been turned down.[112] The problem was that in rejecting such overtures the Faculty risked encouraging by default the growth of specialist examining bodies in the rapidly emerging specialties, keen to establish their professional status with separate training schemes and examinations, which could undermine the whole bid to become a postgraduate teaching and examining body. This tension had to be carefully negotiated for at least the next twenty years. Other diplomas in Psychiatry and Radiology were widely discussed, but temporarily shelved at this point. Nevertheless, Diplomas in Radio-Diagnosis and Radio-Therapy and in Dental Orthopaedics were established during 1948–49, although the latter was rejected as an additional qualification by the GMC.

The problem of the tension between specialist diplomas and the Fellowship as competing postgraduate qualifications was left to simmer on into the 1950s and beyond. However, the Faculty had made sure that the Fellowship, as noted above, would not be left behind as a potential qualification for consultants/specialists, by introducing an extended specialist component as early as 1942. In 1949 this policy was carried further. In fact 1949 was a watershed for the Fellowship, since then, for the first time, a candidate could take an entirely specialist Fellowship in Surgery, without taking General Surgery at all.

The 1940s saw the beginning of the bifurcation of the development of postgraduate medicine and surgery which became a characteristic feature in the second half of the twentieth century. At this stage specialisation was proceeding faster in surgery than in medicine. Consequently, between 1942–46 the Fellowship *qua* Physician was changed making it a more *general* set of examinations. By 1924 the list of selected subjects that a candidate could profess, and take part of the examination in, had grown very long, with a preponderance of medical subjects on offer.[113] As Manderson (et al.) have explained,

---

[110] Ibid. My emphasis.
[111] Ibid., 16 March 1948.
[112] Ibid., 18 May 1948, RCPSG, 1/1/4/7.
[113] See 1924 regulations, RCPSG, 1/5/1/11.

During the second world war criticism of the proliferation of selected subjects mounted within the Faculty. The view was taken that training in a specialty should properly follow training in general medicine and that the option of selected subjects might encourage candidates, without appropriate special training, to regard themselves, nevertheless as 'specialists'.[114]

Thus, from 1942, those wishing to gain the Fellowship *qua* Physician had now to take extra written and oral examinations in the Principles of Medicine (combined with Medical Pathology which was already required in 1924), and in the Practice of Medicine (combined with Therapeutics which was also already required in 1924), in addition to the existing examination in Clinical Medicine.[115] In a further reform of 1946, the possibility of a partly specialist Fellowship *qua* Physician (General Medicine – in its 1942 revision – plus a selected subject) was removed, and the examination then consisted simply of the other 1942 components. This was a departure from the Edinburgh Physicians' model, and prefigured the consensus which emerged in the 1960s on the Fellowship *qua* Physician as a general examination to be taken early in the hospital career, as a prelude to specialist training. This emerging consensus, partly due to the slower development of specialties in medicine, was to lead to the development of a common MRCP in 1972, while the Surgeons are only now (1999) close to achieving this goal.

Conversely, in 1949, the Surgical Fellowship changed to encompass the possibility of an *entirely* specialist Fellowship for the first time. Intending surgical Fellows now had to take a primary examination consisting of written and oral examinations in Anatomy, Physiology, Pathology and Bacteriology, and then a final examination which included written and oral sections in Surgical Anatomy, Surgical Pathology and Surgery, and an examination in Clinical Surgery. However, the new regulations stated that candidates might submit themselves for examination in 'one of the following special subjects in the final examination: Obstetrics and Gynaecology; Ophthalmology; Oto-rhino-laryngology'. As well as the normal conditions for entry to the Final Surgical Examination, such candidates had also to have spent at least one year in an approved hospital clinic devoted to the chosen speciality. The specialties chosen were not controversial emerging ones, but long-standing ones, but the really radical change came in the nature of the examination for such specialist candidates:

---

[114] P. R. Fleming, W. G. Manderson, M. B. Matthews, P. H. Sanderson and J. F. Stokes, 'Evolution of an Examination: MRCP (UK)', *BMJ*, 13 April 1974, pp. 99–107. Quotation from p. 101. This article will be cited hereafter as by Manderson et al. because of Dr Manderson's close association with the Glasgow College.

[115] See *Faculty Minutes*, 7 September 1942, RCPSG, 1/1/1/16.

The Final Examination for such a Candidate shall consist of a written examination, an oral examination and a clinical examination *in the special subject*; and a written examination, an oral examination and a clinical examination *in such Medicine and Surgery as is related to the special subject.*[116]

The Faculty had made the Fellowship *qua* Surgeon respond to changes in medical education and practice, and had enabled it to compete with specialist diplomas, by making it a specialist examination. This move was in direct response to the debate about whether such higher qualifications or specialist diplomas were to be in future the passports to consultant status, a debate that, as we have noted, became very heated in the late 1940s and continued until a consensus emerged at the time of the Todd Report (1968). However, this was as far as the Faculty was prepared to go along the road to a specialist Fellowship. After this concession to prevailing trends, the Fellowship remained a largely generalist qualification.

Building Faculty Status

While the debates about the future nature and scope of postgraduate education were continuing, the Faculty was also engaged in pursuing a range of other moves to boost its flagging public image. Some of these built on previous attempts to become involved in social policy issues, and some were more closely linked to the central postgraduate debate. For example, the revival of plans to unite all the Scottish royal medical corporations into a single Scottish College was inspired by the desire to shore up the status of Scottish qualifications (Fellowships, Memberships and specialist diplomas) against the latest round of threats from London. The desire to stop prospective Fellowship candidates from leaking away into the specialist diploma market, and to link a standardised British Surgical Fellowship intimately with the very definition of a consultant – the grade which was the basis of all the new health care reforms – was also behind successful moves to streamline the structure of the examinations for the surgical Fellowship. These moves resulted in a common British primary examination eventually being achieved.

The Reform of the Surgical Fellowship

From 1944 the Faculty was engaged in a cooperative effort with the other British and Irish surgical colleges to devise a standardised national format for a joint higher surgical qualification. The NHS and the new structure of

[116] *Regulations Respecting the Fellowship of the Royal Faculty of Physicians and Surgeons of Glasgow*, June 1949, RCPSG, 1/5/2/8. (My emphasis.)

medical education and practice were based on the primacy of the consultant grade. If the royal medical corporations were to completely annex the post-graduate education of the consultant, their higher qualifications had to be recognised as the passports to consultant status. Thus moves by specialist groups to institute their own diplomas had to be headed off, and a national structure of consultant training evolved in which the higher qualifications became an integral part of the definition of a consultant. While the debate about whether higher qualifications should be specialist qualifications taken at the end of specialist training as final indications of capacity for consultant status, or at the end of an initial period of general consultant training as an index of potential for further specialist training which might then be tested by some other mechanism later on, continued, it was clearly counterpro-ductive for the royal medical corporations to continue to offer competing higher qualifications. A standardised national scheme would more easily become accepted as a passport to consultant status. Thus from 1944 the various surgical bodies began to meet, to discuss rationalisation plans ema-nating from the umbrella body for surgeons, the Association of Surgeons of Great Britain and Ireland (ASGBI). This effort was part of the remarkably prescient vision of Roy Frew Young for the future of the Faculty. Young's role was not restricted to his work as President in relaunching the Fellowship during 1940–42, noted above. During 1944 he represented Faculty at a series of meetings on the reform of the surgical Fellowship, and convened a special Committee of Surgeons which examined the whole question of the future role of the Faculty, whose conclusions were of enduring value.

Young was a member of the Council of the ASGBI from early 1944. On attending his first meeting of the body he ascertained that it was determined to play an active role in the future training of surgical consultants. With this end in view, it had already been working in close cooperation with the RCSEng in England, and had produced a joint scheme for postgraduate surgical training. Young found, however, that, while the Edinburgh and Irish surgical colleges were recognised by the Association and were fully involved in their plans, the Faculty 'had no standing whatever'. On his own initiative, Young wrote to the association's secretary (General Michener) pointing out the status and position of the Faculty. Michener conferred with the President of the English surgeons (Webb-Johnson), who then issued an invitation to the Faculty to send a representative to an imminent meeting of the English, Edinburgh and Irish surgeons at the RCSEng on 13 January 1944.[117] Young had also written to the president of the association (Sir James Walton) asking

---

[117] Details from R. F. Young, memorandum entitled, 'II. Confidential Report', *Council Minutes*, 1 February 1944, RCPSG, 1/1/3/12.

for the Faculty to be allowed to join in the plans as a full and equal partner. At the next meeting of the Association, Young was requested to expand his evidence for this and he had to clarify a point which had not been clear: that the Faculty elected the great majority of its Fellows in Medicine *or* Surgery only *after examination*,[118] just like the English Colleges. He reported that, 'when it was definitely established that decisions with regard to Surgical Consultants lay with the Surgeons of this Faculty, any objections were withdrawn and the position of the Faculty was unanimously accepted'.[119]

This process of verification of the Faculty's credentials was repeated before the 13 January meeting at the RCSEng. Having been approved, the Faculty was allowed in to be an official part of the scheme. Like James Watson's similar pleading of the Faculty case for full involvement in the 1858 reforms which ushered in the change to national regulation of (undergraduate) medical qualifications, this (in the postgraduate sphere which was now of crucial importance for Faculty survival) was of equivalent import. Young had now assured the Faculty a full place at the top table of postgraduate education. As Young himself put it, his personal efforts had ensured that, 'the position of the Faculty is on a sound and friendly footing with the other surgical bodies'.[120]

At the 13 January meeting at the RCSEng, the association's scheme for future higher qualifications in surgery was outlined. Each royal medical corporation was to institute a two-stage examination. The primary examination would be in Anatomy, Physiology and Pathology and would have to be passed before proceeding to the final clinical examination. With the exception of Ophthalmology (for which 'a knowledge of general surgery is not essential'), there was to be no specialist component: the new joint higher qualification for consultant surgeons would reemphasise the primacy of a general conception of surgery, and would thus be a necessary preliminary to further specialist training.[121] The conference established that the RCSEng was working in close contact with the Minister of Health, and that he evidently (in line with the Goodenough proposals) wished the royal medical corporations to have full control of consultant training. Young commented on this scheme that it would mean the Faculty instituting a preliminary examination in the basic sciences – as the Royal Colleges of Surgeons in

[118] That is excluding the small number of Fellows elected annually on the basis of their clinical or scientific reputations, or published work, and the small number of Honorary Fellows also elected annually.

[119] Young, 'II. Confidential Report'.

[120] Ibid.

[121] Details and quotation from Memorandum by R. F. Young, 'I. Meeting at Royal College of Surgeons, England', *Council Minutes*, 1 February 1944, RCPSG, 1/1/3/12.

England and Ireland had already done, and as Edinburgh was about to do – and that this would necessitate the involvement of the University to provide taught courses. Young recommended that the Faculty appoint a special Committee of Surgeons to conduct negotiations with the other surgical bodies with full Faculty authority. He noted that all this left the position of the Fellowship *qua* Physician undetermined, and recommended the establishment of a Committee of Physicians to conduct exploratory talks with the Edinburgh Physicians with a view to moving towards a common national examination, as was the current trend in Surgery. The Faculty Council responded to this immediately by formally establishing the two committees at the very same meeting at which Young gave his report of developments summarised above, on 1 February 1944. The Committee of Surgeons consisted of Young (Convenor), Scouler Buchanan, G. T. Mowat and William Sewell, and the Committee of Physicians of Charles Fleming (Convenor), McNee, Morris and Snodgrass. The Surgeons were to liaise with the other surgical bodies in moves towards a common surgical Fellowship, the Physicians to open discussions with other bodies on a similar move in Medicine, but both were also to have a wide remit and to discuss, 'all steps to be taken to improve the status of the Faculty'.[122] By the 15 February Council meeting the Committee of Physicians had already produced a unanimous interim report which recommended the formation of a Scottish College of Physicians (the Glasgow Faculty thus becoming a surgical body only) to produce a Scottish Medical Fellowship, as a parallel move to the posited institution of a UK Surgical Fellowship. The Committee felt this was the simplest and quickest way of elevating the status of the Medical Fellowship.[123]

The Committee of Surgeons produced a much fuller report for the 28 March 1944 Council meeting.[124] As well as restating the changes to be made to the Glasgow Surgical Fellowship to bring it into line with the evolving plan for a common surgical Fellowship, this report fitted together the new structure of the surgical Fellowship with a vision of the future of the Faculty as a postgraduate body, in which, in the longer term, a Register of Consultants and Specialists would be kept by the GMC, which would also inspect and sanction postgraduate examinations, as it already sanctioned undergraduate ones. The committee flatly rejected, after a detailed analysis of the Faculty's rights, duties and obligations, the idea of merger into any Scottish College, although it did recommend that for an assured future the Glasgow 'Faculty'

[122] Ibid., 1 February 1944.
[123] Ibid., 15 February 1944.
[124] 'RFPSG Report by the Committee of Surgeons Appointed by the Council', ibid., 14 and 28 March 1944.

would have to become a 'College' to ensure an equal status with the other royal medical corporations. With this change, the committee felt, the Faculty could hold on while the new postgraduate role developed sufficiently to become the professional and financial core activity of the body. This 1944 report from Young's committee thus sketched the entire postgraduate programme for the future of the Faculty.

In December 1944 both of these committees were made standing committees of Council. They are both thus the immediate predecessors of the current Medical and Surgical Advisory Committees.[125] Further negotiations between the surgical bodies occurred during 1945–47, and in May 1947 it was agreed that new joint regulations for the surgical Fellowship would come into effect in 1949.[126] This meant that all UK surgical bodies would have the same basic examination structure to their higher qualifications. This was a necessary preliminary step before reciprocity of examinations could be achieved, and then moves could be made towards a common UK examination. Discussions of reciprocity were still going on in 1949.[127]

Discussions about merger with the Edinburgh Colleges likewise continued until 1949. The merger of the Faculty's Physicians with the Edinburgh Physicians was soon rejected, but other schemes involving the amalgamation of the three corporations into a single or two Scottish Colleges were discussed throughout the latter part of the period covered by this chapter. The motivation behind such moves was to increase the status and thus the attractiveness of Scottish higher qualifications, for it was by no means certain at this time that the Faculty would be able to succeed alone as a postgraduate examining body in competition with London. The advantage of one (or two) Scottish Royal Colleges would not only be increased status south of the Border, but also that graduates in Aberdeen, Dundee and St Andrew's might take the new examinations, instead of taking those of the London Colleges. However, merger plans were finally shelved, once again, in 1950, possibly because the Edinburgh Colleges were not prepared to shoulder the Faculty's debts.[128]

Other Initiatives

Before the end of the war various other practical initiatives had been undertaken to improve the position of the Faculty. In December 1944 a range of

---

[125] Ibid., 19 December 1944.
[126] *Council Minutes*, 20 May 1947, RCPSG, 1/1/4/6.
[127] See, for instance, *Council Minutes*, 21 June 1949, RCPSG, 1/1/4/8.
[128] *Faculty Minutes*, 4 September 1950, RCPSG, 1/1/1/17.

moves were begun. Individual Council members were asked to approach 'influential and wealthy men' in Glasgow about donations to set up an endowment fund for postgraduate activities. Roy Frew Young agreed to produce a brief pamphlet history of the Faculty for use in this campaign. The Council urged Fellows to be sure to advertise their Fellowships in any publication, receptions for representatives of the public, medical practitioners and students were discussed, and plans were made to brighten up the Faculty Hall. There was also a renewed determination to take a more active part in public affairs in order to raise the Faculty's profile. President Sewell, Geoffrey Balmanno Fleming and Stanley Graham began work on a memorandum on the health of Glasgow.[129] This was published in the *Glasgow Herald* on 2 June 1945. It drew attention to the offensively high infant-mortality rate in Glasgow and blamed the high incidence of preventable diseases on social causes, stating that:

> In spite of the fact that 100,000 houses have been built in or in the immediate neighbourhood of Glasgow between the two wars, our housing conditions are still worse than in any of the great cities of the kingdom ... The problem must be faced. Even a vast increase in our hospital and other health services will not solve the problem. We must accept the fact that adverse social and economic conditions are at the root of the matter. It is therefore largely a social problem with which we have to deal. Once the poor are properly housed, are able to earn a wage sufficient to purchase sufficient food and clothing, and are educated in the art of living, there is no doubt that our infantile mortality will come into line with other more advanced communities and the general health of the population will be immensely improved.[130]

The memorandum was just the kind of pro-active intervention in public health matters that Chalmers had attempted to get the Faculty involved in during the early part of the century. It demonstrated the Faculty's interest in a National Health Service that was more than a *sickness* service, providing careers for consultants. Instead the memorandum argued that it should truly be a *health* service, which paid attention to preventative medicine. The memorandum also argued for the coordination of the welfare clinics, the children's hospitals and the University Department of Paediatrics into 'a comprehensive child health service', that anti-tuberculosis efforts should be redoubled with liaison between family doctors, tuberculosis medical officers, health visitors and housing authorities, that hospital maternity provision

---

[129] *Council Minutes*, 4 December 1944.
[130] 'Glasgow's Health Problem: Infant Death Rate Symptom of "Appalling Conditions"', article based on (large quotations from) a memorandum by the RFPSG on the health of Glasgow, *Glasgow Herald*, 2 June 1945.

should be increased urgently.[131] On the back of the success of this memo-randum, both as intervention in social problems, and as a device for raising the public profile of the Faculty, there were discussions in early 1945 about the President convening representative committees to deal with the problems of housing, social services and child welfare, but it is unclear what the fate of this initiative was.[132]

Another important part of moves to improve the position of the Faculty initiated in December 1944 was the introduction of a voluntary annual subscription.[133] By May 1945 this had however only brought in the sum of £281 19s. 0d. (with a further £229 14s. 0d. separately donated).[134] This was not enough to keep the Faculty afloat considering the precarious state of the income from the Fellowship and the financial commitment that was going to have to be made to postgraduate education, a subject that was to return to prominence during the 1950s.

In July 1948 the Faculty joined the other royal medical corporations and the BMA in constituting the Joint Consultants Committee (JCC) to be a unified voice representing consultants and specialists to the government.[135] This underlined the corporations annexation of all postgraduate matters, and the JCC was to become the most influential of the many unofficial medical advisory bodies in contact with the state.[136]

The Faculty celebrated its 350th anniversary in November 1949. It was a celebration of the survival of the Faculty as a meaningful licensing body and a confirmation of the hope that postgraduate education would prove to be the new mainstay of its professional role for the future. As we celebrate the Royal College's 400th anniversary we know how well placed this hope was. But in 1949 the future was still far from certain. As the Honorary Treasurer reported to Fellows with sobering effect at the November 1949 AGM:

> At first glance the position seems highly satisfactory in that we have ended the year with a considerable credit balance – namely £2413 5s. 3d.: in reality, the position is most disquieting, and I would ask your most careful attention to the following remarks ... We derive our income mainly from three sources: 1. Fees from Candidates; 2. Subscriptions [still voluntary at this point]; and 3. Interest on Capital ... In other words, nine-tenths of our income is derived from exam-ination and entry fees for the Fellowship, the Triple Qualification and the Dental

[131] Ibid.

[132] *Council Minutes*, 27 February 1945, RCPSG, 1/1/3/13.

[133] Ibid., 4 December 1944.

[134] Ibid., 15 May 1945.

[135] Ibid., 21 September 1948.

[136] For more on the JCC, see Rosemary Stevens, *Medical Practice in Modern England: The Impact of Specialization and State Medicine* (New Haven, 1966), passim.

Licence. The fees from the Dental students are down by about 50 per cent from last year, and will soon cease entirely. The fees for the Triple Qualification will likewise disappear. The special conditions for the Fellowship no longer exist. The post-war flow is ebbing fast. The new Regulations for the Surgical Fellowship are already exerting a strong deterrent effect on potential candidates, and I think it is not unfair to say that the administration of the new Health Service has had a most unsettling effect upon young graduates. The net result of all these factors, Gentlemen, will be to reduce the nine-tenths of our income to a very small fraction indeed.[137]

And this was before expenditure on postgraduate education was taken into account.

It is thus clear that the practical policy of cooperation with the University, and particularly the annual grant of £1000, was crucial to the Faculty's survival while it reoriented itself towards a role in postgraduate education and examination. While there were still areas of dispute between the Faculty and the University, for instance in regard to representation of the Medical Education and Appointments Committees of the new Western Regional Hospital Board (WRHB),[138] the close connection with the University, and particularly with Sir Hector Hetherington was extremely important. It should come as little surprise then that, in the anniversary year, Hetherington was elected as an Honorary Fellow of the Faculty, and his portrait appears with those of the past Presidents in Tom Gibson's useful short history of 1983.

The period from 1944 was the key one in the history of the Royal Faculty in the twentieth century. A new symbiotic relationship with the University, now the focus of the Glasgow medical scene, opened the way to a new mission. As Hetherington said in that other anniversary year, fifty years ago, when accepting his Honorary Fellowship:

It is a matter of history that the relations between the Royal Faculty and the University have not always been of the easiest – there were battles long ago. But for many years now the two Societies have worked in growing harmony, and, as I think, to the greater good of both and of the community which both exist to serve ... We have in large part a common membership, and we have completely a common task. I hope we shall long walk together in unity of spirit and of purpose.[139]

[137] Honorary Treasurer's Report and Financial Statement, *Faculty Minutes*, 7 November 1949 AGM.
[138] See for instance the exchange of letters reproduced in *Council Minutes*, 18 May 1948, RCPSG, 1/1/4/7.
[139] Sir Hector Hetherington, speech to RFPSG Anniversary Dinner, 29 November 1949, GUABRC, Hetherington Papers, DC8/137.

As the Faculty celebrated its 350th Anniversary in 1949, however, the task of reorienting itself towards a new future as a primarily postgraduate teaching and examining body was only just beginning. The writing on the wall had been read, to be sure, but there was little flesh on the dry bones of the idea of the Faculty as a postgraduate body. Many of the Faculty's old guard still doubted whether the transformation could be accomplished, or was even desirable. Certainly postgraduate education and teaching was so new that even the demand for it had yet to fully emerge. There was now only the realisation that this was the only way forward. Financial returns from postgraduate education were still a long way off.

While Alstead acknowledged the pivotal and pioneering role of the remarkable strategic foresight of Roy Frew Young in these years both as President at the beginning of the period, and dedicated committee member later on, which we have outlined in detail above, this, I think is what he had in mind when he criticised Tom Gibson for so forcefully locating the 'Turn of the Ebb' tide in Faculty affairs during Young's Presidency. Notwithstanding all Young's efforts, Alstead commented,

> It is probably true to say – as Mr Gibson says – that the prospect of new developments ['the Government ... displaying interest in the future of the country's medical services'] enabled the President to infuse optimism into the Fellows who attended the Faculty meetings. Thus he achieved largely by dint of his strong personality, transparent honesty, courage, and – not the least – because he enjoyed a high reputation as a surgeon and teacher. *Nevertheless, he could not materially change the financial affairs of the Faculty.*[140]

Readers may remember Alstead's parody of the Glasgow College as the poverty-stricken 'Royal College of Physicians and Surgeons of Chowbent'. If there isn't enough money to pay the bills, fictional and real medical corporations are doomed. Now that the way ahead was clearly postgraduate education and examining, with University support instead of opposition, it was left to a new generation of younger men to flesh out this project and, while the Faculty established itself as a postgraduate body, to put it on a firm financial foundation.

---

[140] Stanley Alstead, 'Preface' to his *Extracts*, p. xxvii, my emphasis. See also letter from Alstead to Tom Gibson, 19 September 1983, a copy of which is in the Shaw Papers, RCPSG, 44. Here Alstead notes that Young made a, 'lasting and favourable impression on representatives of the other medical corporations ... when the Faculty was struggling to recover status and dignity', and also that 'the prospect of our having an NHS galvanised some local specialists into taking the Fellowship ... but of course, this well soon ran dry'.

Facultas Regia Medicorum et Chirurgorum Glasguensis

REGIO MUNERE INSTITUTA:

Videlicet per Chartam Jacobi Sexti A.D. MDXCIX. Concessam, et Acto Parliamenti Carolo Secundo regnante, sancitam; necnon Acto Parliamenti Anno decimo tertio Victoriæ Britanniarum Reginæ, Cap. XX, explicatam et emendatam.

Lectoribus Salutem.

Sciatis omnes,

VIRUM PERQUAM GENEROSUM

Andrew Watt Kay, M.D., Ch.M

Curriculo Disciplinarum ad Medicinam ac Chirurgiam Spectantium rite peracto, et scientia eius nostro examine tam publico quam privato luculenter probata, in Sociorum Facultatis Regiæ Medicorum et Chirurgorum Glasguensis numerum adscitum esse.

In cujus rei testimonium Præses noster, Inspector, Quæstorque hisce literis Sigillo Communi munitis nomina sua manibus propriis apposuerunt.

Ex Aula nostra apud Glasguam A.S. millesimo nongentesimo quinquagesimo sexto mensis Februarii die sexto

PRÆSES.

INSPECTOR.

QUÆSTOR.

Fellowship *qua* Surgeon Diploma of Sir Andrew Watt Kay, 6 February 1956.

4

Postgraduate Teaching and Examining,
1950–1960

I confess once again that I have a limited interest in the past … In the twentieth century this Medical Corporation cannot be run on a basis of hero worship and by resorting to Victorian concepts of domestic economy. We must be constantly adapting ourselves so that we may accurately interpret the spirit of the times in which we live. Let me come nearer home. I suggest to you that the Faculty can make its greatest contribution to the future of medicine in Scotland by showing its determination to take part in shaping activities in all kinds of Postgraduate Medical Education. This can certainly be achieved by the combined efforts of the Hospital Staffs to which we belong. If we are successful, certain important developments will follow as surely as day follows night … or we shall be left stranded, remote from the main stream of medical thought and action … we shall see to it that the administration and the amenities of the Faculty Hall are provided for in a way that is realistic and in keeping with the life of a great Industrial City.[1]

In the 1950s the Faculty completed its transformation into a postgraduate teaching and examining body.[2] This reorientation had been forced on it, as we noted in the previous chapter by the Goodenough and NHS changes which had finally given the University sole regional control of undergraduate education, and had reinforced the long emerging professional division between GPs who worked in the community and consultants/specialists who worked in hospitals. However, while undergraduate education had been effectively standardised, postgraduate education, on which the whole edifice of interlinked reform of medical education and provision ultimately rested because of the centrality of the highly-trained consultant to the NHS, had been left in an archaic state of *laissez-faire* and *Lernfreiheit,* which had previously characterised undergraduate qualifications before the state and

---

[1] Stanley Alstead, PRFPSG, addressing the 3 November 1958 AGM, *Faculty Minutes,* 3 November 1958, RCPSG, 1/1/4/17.

[2] The main basic sources for this period are Alstead, *Extracts,* and Gavin Shaw, 'RCPSG'.

the universities had completed the long process of monopolistic standardisation in 1948. Goodenough had recommended that postgraduate qualifications be left to the control of the royal medical corporations, but no fixed
definition of the consultant had been laid down. There was no GMC control
of curricula, no standard courses of instruction and no guidelines as to which
qualifications were most suitable as passports to the key consultant status.
Potential consultants had to shop around for the most prestigious qualification, and for courses of instruction leading to it. In this context the Faculty
had to make the Fellowship as attractive as possible to potential candidates.

The Faculty had been reviving its main existing postgraduate qualification,
the Fellowship *qua* Physician or Surgeon, during the late 1940s to become a
passport to consultant status, grasping that postgraduate qualifications were
the key to its continued relevance. After the end of the Second World War
a large number of demobilised doctors returned to Glasgow, after intense
and varied wartime medical experiences, wishing to complete their training
and gain professional advancement on the new career ladder to consultant
status. The end of the war also witnessed the beginning of a continuing
stream of foreign trainees (initially often from the ex-colonies) into Britain
looking for hospital placements and training for prestigious British qualifications. Demand was therefore present, and it was up to the Faculty officers
to revise and represent their product (the Fellowship) as a prestigious entree
to the consultant cadre. The battle to get the Faculty's Fellowship recognised
as on a par with those of Edinburgh and London was a continuing problem,
but new strategies were now tried. In January 1951 reciprocity of the Primary
Surgical Fellowship examinations was agreed between the Faculty, the
RCSEng, the RCSEd and the RCSI (the Royal Australasian College of Surgeons entered the accord in 1952 and the South African Surgeons later also
joined).[3] Candidates took a common primary exam and could later choose
which body's final exam to take, although in 1959 moves were also begun
towards a common final exam.[4] The Faculty's Surgical Fellowship gained
status by being thus bracketed with the London and Edinburgh examinations,
and competition was replaced by symbiotic cooperation. This was also the
period when (ultimately fruitless) attempts were made to form a joint diploma awarding body of the Scottish universities and royal medical
corporations. Again the aim was to offer credible local alternatives to London
qualifications.[5]

The Faculty's relationship with the University Medical School had under-

[3] *Council Minutes*, 16 January 1951, RCPSG, 1/1/4/10.
[4] Ibid., 21 July 1959, RCPSG, 1/1/4/18.
[5] Ibid., 20 February 1951 to 19 February 1952, passim.

gone a sea-change: the two bodies no longer competed for the same students, and were increasingly made up of the same people, pulling in the same direction. Hetherington had revitalised the Glasgow Medical School with a series of progressive appointments and an extension of the academic ethos of interlinked teaching and research into the hospitals via clinical units which encouraged the growth of specialty and clinical research. Hetherington's men came to play an important role in the direction of the Faculty in this period. They were all actively concerned to establish a postgraduate *school* in Glasgow to provide trained personnel to staff their units. From 1952, for example, via the forum of the GPGMEC (a University Committee on which the Faculty was represented) the University ran two courses preparing candidates for the Primary Surgical Fellowship, one of which was at the University's Anatomy Department. In addition, the University's £1000 grant to the Faculty for developing postgraduate education was continued throughout the 1950s, serving as a potent reminder that the University was now the local educational power-centre, while the Faculty was developing as its postgraduate teaching and examining arm to prepare consultants, partly for the academic units in the hospitals. This background support for progressive reform was actively utilised by a reforming triumvirate of Faculty officers: Stanley Alstead,[6] Archibald Goodall[7] and Gavin Shaw.[8] Between them, and with some good

---

[6] Stanley Alstead (1905–1992): He graduated MB, ChB from Liverpool University in 1929. After junior hospital posts in Liverpool, Birmingham and Salford, he obtained his MD from Liverpool in 1931 with a thesis on heat in diphtheria using the electrocardiogram. He was appointed Robert Pollok Lecturer in Materia Medica and Pharmacology in the University of Glasgow in 1932, with clinical responsibilities in the GWI. He became a Fellow in 1935. He moved his clinical base to Stobhill when Morris succeeded Stockman in 1936, and succeeded Morris in 1948. In an era when Materia Medica was becoming Clinical Pharmacology, he viewed the patient as a whole, not just the drug and 'struck a nice balance between the traditional and the modern'. Like Morris, he pioneered the specialty of geriatrics. He served on the Council of the Faculty, was Visitor from 1954–56 and President from 1956–58. His obituarist also noted that, 'his was a key influence in modernising its policies and its premises, in promoting and publicising its activities, and particularly in developing its role in postgraduate medical education'. Obituary by Professor Edward McGirr, *Glasgow University Year Book* (1993), pp. 71–74.

[7] Archibald Lamont Goodall (1915–1963). Graduated in Medicine from the University of Glasgow in 1937. He became attached to the University Department of Surgery in the GRI under Professors Burton and Mackey. He became a Fellow (*qua* Physician) in 1940, and he took his MD at Glasgow in 1944. He attained the rank of Consultant Surgeon in 1948, and held an Honorary Lectureship in Surgery in the University. From 1946 he was Honorary Librarian, and was thus a member of Faculty Council for seventeen years. He was vice-convenor of the GPGMEC, and later adviser to the Postgraduate Medical Board. He was a keen medical historian and founder member of the Scottish Society of the History of Medicine.

[8] Gavin Brown Shaw (b. 1919). He graduated from Glasgow University BSc (1939) and MB,

luck in the form of financial gifts and bequests, these three set the Faculty on course as a postgraduate teaching and examining body with an integrated system of courses, facilities and examinations, and a rationalised administrative structure.

On 14 May 1951 a special Faculty meeting was held as a forum for discussion of the linked themes of the finances and future role of the Faculty as a medical corporation. The Honorary Treasurer, Matthew White, identified the central problem, stating that, 'in a very short time all income from candidates for the Licence [in Medicine and Surgery, i.e. the TQ] and the Licence in Dental Surgery would cease and the Faculty would require to depend on income from candidates for the Fellowship and from invested funds'.[9] The end of the extra-mural schools and the University's take-over of undergraduate education, White judged, would lead to an annual deficit of about £3000. At this meeting the Faculty appointed a committee of younger Fellows to consider the idea of a voluntary subscription from all Fellows, and a 'clear policy' for the future.[10] This committee (which included Shaw, A. B. Kerr, Eric Oastler and George Henderson Stevenson) reported to the Council in June and argued for a voluntary subscription and recommended, 'in the most urgent terms, that the Faculty should take a more active part in Post Graduate Medical Education'.[11] It suggested that this should be done by forming a Faculty Postgraduate Society open to all medical practitioners; by extending the existing Faculty lecture series; by initiating sectional meetings; by turning the library into a study centre; and by making the Faculty a local medical information centre. The Council appointed a sub-committee to consider these recommendations which included Thomas Anderson, Stanley Alstead, Archibald Goodall and George Wishart. Reporting in December, this committee (which had also consulted with representatives of the RMCSG) also came out in support of a voluntary levy on Fellows and of a reorientation towards postgraduate education. It was uncertain how to proceed on this

ChB (1942). After war service, he was Clinical Tutor to Professor McNee at the GWI until 1948 when he was appointed Senior Registrar in Medicine at the Southern General, where he progressed to Consultant Physician with duties in Cardiology in 1956, and to Consultant Physician and Cardiologist in 1963. He retired from this post in 1984. He became a Fellow *qua* Physician in 1950, and was Council Member from 1953–57, when he became Honorary Secretary until 1965. He returned to the Council from 1970–74, was Visitor from 1976–78 and President of the RCPSG from 1978–80. He was a member of the GMC from 1981, and was also awarded the CBE in that year.

[9] *Faculty Minutes*, 14 May 1951, 28 January 1952.

[10] *Council Minutes*, 26 June 1951 (insert), 'Report by the Committee Appointed by the Faculty on Monday, 14 May 1951'.

[11] Ibid.

latter point, noting only that the future of the Faculty was closely linked with other bodies: the University, the RMCSG, and the other Scottish corporations. The committee favoured as much cooperation as possible between the medical corporations short of merger, including exchange of examiners. It was wary of antagonising the RMCSG, which was a long-standing forum for academic papers and lectures, and it felt that the Faculty should clarify its sphere of influence in postgraduate education with the University, arguing that the Faculty, 'must be clear as to its responsibilities in postgraduate education and the part it should play'.[12]

Progress was made in one of these areas when by the end of 1951 a permanent Standing Joint Committee (SJC) of the three Scottish royal medical corporations was established to speak with an authoritative voice on all questions of mutual concern. The SJC became an increasingly important forum for establishing joint lines of action, The remaining points were to concern the Faculty for the remainder of the 1950s and beyond.

In January 1952 another special meeting of Faculty was called to discuss future policy and specifically a memorandum from the President (Walter W. Galbraith, a pioneering urological consultant surgeon and chief at the GWI) to all Fellows which summarised the deliberations of the above committees and recommended graded voluntary subscriptions until powers could be obtained to introduce a mandatory levy (a policy being considered by all the medical corporations), and the appointment of a permanent Council sub-committee on the future direction of the Faculty which would evolve a long term policy in tandem with the University, the other corporations and the RMCSG. All contributors to the meeting agreed that postgraduate education was the key to the future, but there were still a range of views on how best to proceed. A memorandum from the ailing Geoffrey Balmanno Fleming (the Emeritus Professor of Paediatrics at the University and Honorary Consulting Physician at the Royal Hospital for Sick Children) was read in which he urged, 'that the salvation of the Faculty rested in the development, in close association with the University, of a *strong and active postgraduate medical school in Glasgow*'.[13] This could function as a local training ground for scientifically trained consultants to staff the University and the hospitals, as the British Postgraduate Medical School at Hammersmith had been doing for the London Medical Schools since its establishment in 1935. The Visitor, Andrew Allison, noted that a stage in the evolution of medical qualifications had been reached, 'when many of the functions of the Faculty have been

---

[12] *Council Minutes*, 18 December 1951 (insert), 'Report by the Committee set up by the Council to Consider the Memorandum Presented by the Committee of Fellows'.
[13] My emphasis.

usurped by the Government, and the University'. William Reid (surgeon at the GRI) suggested that the Faculty should 'throw its doors open to all medical practitioners in Glasgow and the West of Scotland' with more up-to-date lecture topics and weekly sectional meetings. Arthur Jacobs (a pioneer of the specialty of urology, who had attained control of his own Department of Urology, the first in Scotland, at the GRI in 1936) summed up the view of the younger Fellows when he commented that he:

> considered that the future of the Faculty was closely associated with postgraduate education – not necessarily with the ultimate aim of candidates appearing for examination for diplomas. He was of the opinion that the Faculty's financial difficulties would disappear if it were successfully to develop postgraduate medical education in Glasgow'.[14]

Jacobs thus shared Balmanno Fleming's view that becoming an effective teaching (as well as examining) body in postgraduate education was crucial to the Faculty's future. The meeting ended with the Faculty remitting to the Council to institute an appeal for voluntary subscriptions, and to establish a Committee to direct 'all aspects of future policy'.[15] This included the Fellowship; all specialist diplomas; postgraduate education; the putative Faculty Postgraduate Society; the use of Faculty buildings; voluntary contributions and finance and 'any other projects concerned with the welfare of the Royal Faculty'.[16]

This Council 'Policy Committee' was set up in February and consisted of Allison, Alstead and W. Ian Gordon. By the end of March the Policy Committee had decided to concentrate its efforts on postgraduate education (it is often known as the Postgraduate Medical Education Committee in subsequent minutes). As the Committee stated, 'It was considered that Postgraduate Medical Education was the most important aspect of the remit, and ... that Postgraduate Medical Education for the Fellowship and the Fellowship and the Finances of the Faculty are closely interrelated'.[17] Meeting in April, with the addition of Illingworth, Oastler, G. H. Stevenson and Wishart, the Policy Committee agreed certain ground rules for setting up courses 'to serve principally those proceeding to higher qualifications'. The courses should combine systematic teaching (lectures) and clinical work, should ideally be synchronised with diets of the Fellowship examination, and the WRHB should be approached to request study leave for Registrars to take

---

[14] All quotations in this section are taken from *Faculty Minutes*, 28 January 1952, RCPSG, 1/1/4/11.

[15] *Council Minutes*, 19 February 1952.

[16] Ibid.

[17] Minutes of the Postgraduate Education Committee, 25 March 1952, RCPSG, 1/1/4/11.

the courses. Illingworth was to head a small committee to draw up a surgical course, with Alstead doing the same for medicine.[18] The courses were to be 'predominantly clinical' and would be taught mainly by the younger hospital staff,[19] most of whom were very keen to become involved in teaching as a way of preparing for entry to the academic stream of medicine as members of university units.[20] The bias of the clinical teaching was to be towards exposing candidates to a wide range of clinical material and encouraging maximum participation so that the course would act as an 'intensive refresher course in bedside diagnosis and clinical discussion'.[21] So, while the Faculty was marking its commitment to the new ethos of specialisation by introducing such courses, its traditional commitment to the importance of general medicine and surgery and the critical importance of the skills honed in the clinical encounter as the basis of all medical knowledge was simultaneously reemphasised. Practice and wide experience, rather than specialised scientific theory, were still fundamental to the Faculty's conception of medicine: while some Fellowship candidates would go on to become academic consultants, concerned primarily with teaching and research, a greater number would be NHS consultants primarily concerned with patient care. In any case, the Faculty believed that wide clinical experience was the foundation of both types of medicine. As the memorandum concluded, the course description should stress that it, 'would not be allowed to degenerate into a series of formal lectures; and the exclusion of this kind of hackneyed textbook teaching would become the hallmark of the Glasgow method'.[22] This same position, that medical education should include experience of the broad range of medical and surgical practice, was also the theme of Thomas Anderson et al.'s 'Memorandum on the Training of Physicians and Surgeons' which was sent to the both the UK and Scottish branches of the Joint Committee of Consultants and Specialists, to the Central Committee of Consultants and Specialists (Scotland) and to both the Edinburgh Colleges with the full endorsement of the Council in March 1952. The complaint here was that there were problems with the NHS training ladder, and especially with the post of Registrar. This grade had become the stepping stone to consultant status, with the result that once they had obtained it in a particular unit,

---

[18] Minutes of a Meeting of the Policy Committee, 10 April 1952, RCPSG, 1/1/4/11.

[19] 'Policy Committee: Development of Postgraduate Education, Memorandum for Discussion' (undated but March–April 1952).

[20] Interview by the author with Gavin Shaw at the RCPSG, 29 October 1997. I am grateful to Mr James Beaton, the College Librarian, for setting up this interview.

[21] Policy Committee, 'Development of Postgraduate Education'.

[22] Ibid.

budding consultants were reluctant to leave it, even if this meant rejecting opportunities in other posts to broaden their clinical experience:

> As a result, few of the future consultants in paediatrics have acquired experience in infectious diseases, tuberculosis, skin, ophthalmology and neurology and, on the surgical side, of orthopaedics, chest surgery, ear, nose and throat surgery, or genito-urinary surgery. It is also noteworthy that it is now unusual for experience to be gained in laboratory practice, e.g. in pathology, bacteriology, biochemistry, anatomy and physiology [and] for the young physician in general practice.[23]

The suggested remedy was that all posts after qualification should offer training and the prospect of consultant status, not just the coveted Registrar grade, and that posts should be held for one year only (with the possibility of renewal). This would encourage movement and a gradual accretion of medical and surgical knowledge through experience. It would also, 'reopen the specialties as subjects in which short periods of experience were desirable'.[24] The Faculty supported science and specialisation, but only within a unified model of medicine based around the idea of a general body of medical and surgical knowledge, best learned via direct experience.

Action was also being taken to improve the Faculty's finances. In March 1952 a printed appeal went out to all Fellows asking for voluntary subscriptions. This had raised nearly £940 by the time of the AGM in November, though less than half of the Fellows in Glasgow and the West of Scotland had responded, and donations fell off dramatically in the next few years as the novelty of the special appeal waned.[25] The financial position was further eased in September when the Faculty was officially recognised by the Chief Inspector of Taxes as a charity under the Income Tax Acts.[26] The Faculty had sought such recognition after observing the example of the success of the Royal College of Surgeons of England in obtaining charity status earlier in the year. It meant that tax could be reclaimed on voluntary subscriptions made under deed of covenant (and one eye here was clearly on future compulsory subscriptions),[27] and tax could be reclaimed on income from property and investments. In all this added about £1000 to

---

[23] Thomas Anderson with Goodall, Illingworth and Thomas Semple, 'RFPSG Memorandum on Training of Physicians and Surgeons', 15 July 1952, insert with *Council Minutes*, 15 July 1952.

[24] Ibid.

[25] Of the 350 Fellows in Glasgow and the West of Scotland, 143 responded, *Faculty Minutes*, AGM, 3 November 1952, Honorary Treasurer's Report.

[26] *Council Minutes*, 16 September 1952.

[27] Acting swiftly to gain maximum benefit from the new situation, forms for requesting deed of covenant status for subscriptions were sent out to all Fellows after the 16 December Council meeting.

annual income,[28] but with the penalty that the Faculty was not supposed to participate in medico-political activities. However new bodies, for instance the Joint Consultants Committee, on which the Faculty was represented were increasingly taking over such trades union style functions. The position was further strengthened by the donation of £500 and 1500 shares in Metal Industries Ltd. to top up the Barclay Ness Library and Fabric Fund. This fund had been established in 1932 by a donation from Barclay Ness, President from 1931 to 1933, the interest from the initial sum being used for refurbishment purposes.

There were two important developments during 1952 which linked the themes of financial security and postgraduate education. On 7 February the Court of Session had approved the transfer of the old Lock Hospital for Venereal Diseases' funds to the Faculty. This hospital which had closed with the advent of the NHS and the declining incidence of (now) controllable venereal diseases had bequeathed funds totalling about £90,000 to the Faculty, ostensibly for research and public education on venerology. The Faculty clearly needed these funds for the refurbishment of its facilities (especially the library) for postgraduate education. These funds were to be the subject of many legal manoeuvrings to free them up in the years ahead. In 1954, by the legal instrument of *cy-près*,[29] the Faculty obtained the right to use the funds more generally if it was felt that no useful contribution to venerology could be made. In April 1952 the Faculty received a more immediately useful gift: a posthumous donation of £6000 from Geoffrey Balmanno Fleming to be held in trust as a postgraduate education fund, the interest from which was to be used to pay for the establishment of courses. Fleming, whose insight that postgraduate education was the future we noted earlier, could not have better served his Faculty than by this timely gift which was used during the 1950s to begin to establish Glasgow as a reputable teaching and examining centre.

In January 1954 the President, Andrew Allison asked for suggestions for postgraduate initiatives and this elicited a letter from Leslie J. Davis, a Fellow and Professor of Medicine at the University Unit at the GRI, who had been appointed in 1945 as part of Hetherington's modernising strategy for the

---

[28] *Faculty Minutes*, AGM, 3 November 1952, Honorary Treasurer's Report.

[29] Stroud's *Judicial Dictionary* defines it thus: 'The *cy-près* doctrine is one of construction, and this is: where there is a gift or trust for a charity which can be substantially but not literally fulfilled, it will be effectuated by moulding it so that as nearly as practicable the intention of the benefactor may be carried out', p. 612. I am grateful to Catriona MacDonald of the Wellcome Unit for the History of Medicine, University of Glasgow, for digging out this reference.

University Medical School. Davis suggested a Faculty lecture course of twenty or so lectures on themes in medicine and surgery delivered by local hospital staff. Davis did not want this series to be considered as a 'cram course' for higher examinations but as a contribution to the ongoing training of would-be consultants and therefore fulfilling 'a genuine educational function'. He wrote that:

> However assiduously a postgraduate student applies himself to his reading, I believe that he cannot fail to gain much help from good lectures on various aspects of medicine given by people who can speak from experience and with authority on their various themes. Obviously such a course of lectures would not in any way replace the need for regular clinical work for those aiming at higher qualifications, but a large proportion of such candidates in this area already have their hospital attachments. For those who have not, such as candidates from overseas, steps should be taken to help them to obtain appointments as clinical assistants.[30]

This letter drew strong support from Gavin Shaw, at that time Assistant Physician at the Southern General with a developing interest in cardiology. Shaw had been asked to stand for election to the Faculty Council by a group of the younger Fellows who had their own 'Senior Registrars Committee' as their spokesman,[31] and he was duly elected onto the Council in 1953. The younger Fellows were the force driving the Faculty towards its new post-graduate role. They felt 'quite strongly that the City of Glasgow should take a more active part in postgraduate educational activities'.[32] They needed local postgraduate teaching to enable them to gain the higher qualifications now necessary to advance up the NHS training ladder towards consultant status. As Shaw recently commented, the younger Fellows were influential not only because they were the future of the Faculty but also because:

> We were noisy, we made a noise ... We did say the sort of things we wanted: we wanted formal postgraduate education, we wanted a lecture theatre, and we wanted the Faculty to be gradually recognised against Edinburgh and London, who were forging ahead while we were just ticking over.[33]

In Council they had the support of Archibald Goodall, who, as Honorary Librarian, had a *de facto* permanent place on the Council and was not susceptible to the vagaries of yearly reelection, and of Stanley Alstead, who became Visitor (an office that was effectively President-elect) in 1954. Shaw

---

[30] Short and long quotations here from a letter from L. J. Davis to Andrew Allison, 24 February 1954, Gavin Shaw Papers, RCPSG, 44, file 'Postgraduate Education I'.

[31] Interview by Andrew Hull with Gavin Shaw, 29 October 1997, RCPSG.

[32] Shaw to Allison, 16 April 1954, Shaw Papers.

[33] Interview with Gavin Shaw 29 October 1997, RCPSG.

wrote to Allison that Davis's suggestions should immediately be put into effect to establish postgraduate education in Glasgow, arguing that, 'such an activity would be important not only for the prestige of the Glasgow Medical School but also would offer to those younger Fellows who question the purpose of the Faculty's existence, a very real and very live reason for its being'.[34] Shaw suggested informal meetings of young Fellows with similar interests from which lecturers could be drawn for a postgraduate lecture series to be held before each diet of examinations (unlike Davis, Shaw wanted the lectures to be a formal preparation for the examinations). Complementary clinical assistantships in the professorial units should be organised by the Dean of the Faculty of Medicine and Director of Postgraduate Education (Wishart). But while the University would necessarily be involved, the Faculty should take the lead:

> In all these activities, the Faculty Postgraduate Committee would remain the central body for organisation ... it could have an Information Bureau to state where and when different subjects might be studied in Glasgow; what specialised clinics are held; who is carrying out certain operations and techniques; meetings at which graduates are welcome etc.[35]

Both of these letters were discussed at the Council meeting of 20 April and separate committees were appointed to take responsibility for medical and surgical postgraduate education. Shaw was the convenor of the former, and his fellow members were Davis and Sam Lazarus;[36] Goodall was convenor of the latter which also included Walter Galbraith and A. B. Kerr. Shaw realised that this was a long-term policy and that there might be only limited success for many years, and even a loss, before the Faculty was established as a postgraduate centre. He thought the ideal scheme on the medical side would be rotating clinical assistantships in medical wards so that the candidate could

[34] Shaw to Allison, 16 April 1954.

[35] Ibid.

[36] Dr Gavin Shaw commented in January 1999: 'Lazarus was one of four young Physicians sent in the War years from the Western and Royal Infirmaries to the Southern General Hospital, then run by the Corporation's Health Department. The hope was that they would improve clinical standards to the point that, eventually, the University could send undergraduate students for clinical education, just as had happened at Stobhill. The other young Physicians were Laurence Scott and Tom Fraser from the Western Infirmary (as was Lazarus), and Alex Brown from the Royal. At the end of the War, all except Scott returned to their base hospitals, but Scott stayed on and was joined by Eric Oastler. At a much later date, Lazarus and Brown returned as Senior Clinicians. Undergraduate teaching at SGH began in the early 1950s'. Personal communication between Dr Shaw and Andrew Hull, January 1999. Before returning to SGH, Brown had become part-time chief of a non-professorial unit at the GWI in general medicine, with a special interest in haematology, see MacQueen and Kerr, *The Western Infirmary*, pp. 130, 132.

gain experience of different specialist academic units, coupled with four or five weekly lectures on recent advances in medicine and including such areas as therapeutics, radiotherapy and medico-physics. All courses should be of suitable standard to prepare candidates for the Fellowship. Shaw also understood that the Faculty must get the 'blessing' of the University's Postgraduate Medical Education Committee for any courses.[37] By May a provisional list of lectures and lecturers for the medical course to run in winter 1954–55 was complete. Writing to Goodall, Shaw reiterated that they were 'designed for the hospital specialist in training and wishing to sit his Fellowship or Membership examination'. There were to be twelve lectures in a course and there was, experimentally, to be no fee for auditor or lecturer.[38]

Both Shaw's and Goodall's course proposals were approved by the Council in July and the courses ran for the first time from October 1954 to March 1955. At first attendance was encouraging: forty-one practitioners had enrolled by November and an accompanying series of demonstrations had been arranged in Lecture Room B at Faculty Hall, the first on mitral stenosis and patent ductus arteriosus, and the second in late 1955 on the laboratory diagnosis of virus infections.[39] The lecture courses continued in 1955–56, but by this time it had been found necessary to charge a fee of £3 3s. 0d. for non-Fellows to cover costs.[40]

An integrated postgraduate education scheme was now coming together. As well as the postgraduate lecture series, the regular (endowed) Faculty lectures continued and the meetings of the RMCSG also provided a useful postgraduate forum. In 1955 Arthur Jacobs ran a course on urology from 18–20 November which attracted twenty-four participants. He believed there was a need for more such specialised courses on definite subjects, and Goodall (one of the Faculty representatives) agreed to bring this up before the

[37] Details of Shaw's views here from his private memorandum, 'Notes on Postgraduate Education', 28 April 1954, Shaw Papers, in which he set down his views probably as part of the process of drawing up a plan for the medical course with Davis and Lazarus.

[38] Shaw to Goodall, 11 May 1954, Shaw Papers. The provisional list of lecturers included Illingworth on gastroenterology (peptic ulcer); under endocrinology, Oastler on adrenals and pituitary problems and Edward Johnson Wayne on thyrotoxicosis; under cardiovascular disease, J. H. Wright on assessment of patients for cardiac surgery and Shaw on management of hypertension; under therapeutics, Alstead on clinical assessment of therapeutic agents and Anderson on antibiotics and chemotherapy; under neurology, Sloan Robertson on disc problems; and under haematology, Davis on megaloblastic anaemias. Note that most of these lecturers were drawn form the new young scientifically-minded and specialised group of hospital staff which had either been appointed by Hetherington or were products of his appointees' units.

[39] See Council Minutes, 19 October and 16 November 1954, 19 December 1955.

[40] Ibid., 19 July 1955, RCPSG, 1/1/4/14.

GPGMEC. For its part, the GPGMEC was happy to allow the Faculty to make the running in postgraduate education: the committee, which as we noted had been active immediately after the war in arranging clinical attachments to hospitals for returning doctors and ran a course in medicine for higher degrees, was largely moribund. It offered only courses in dermatology, chemotherapy and a preparatory course for the Primary Surgical Fellowship. Meanwhile the Edinburgh Postgraduate Board (a University/RCPEd/RCSEd body that had replaced the Postgraduate Executive Committee in 1945) was still running its highly successful thrice-yearly intensive Internal Medicine course, and had introduced a new twice-yearly surgical course, both of which prepared candidates for the higher qualifications of the Edinburgh Colleges.[41] It was this inactivity in Glasgow which had spurred the Faculty to begin to organise postgraduate teaching itself. Behind the scenes, it is probable that the University was happy to allow the Faculty to proceed, because it had more than enough to do with undergraduate education. The University continued to give the £1000 annual grant to the Faculty for postgraduate activities, and it is likely that Hetherington envisaged the Faculty, as the natural focus for the non-university unit clinical teachers in the hospitals, to act as the University's postgraduate teaching arm. However, the Faculty was aware that there was much loyalty to the University Committee from clinical teachers (whereas in Edinburgh there were a variety of unofficial postgraduate schemes). Therefore the Faculty began to launch itself as a postgraduate teaching body and it was careful to keep the GPGMEC informed at all points and to secure its approval.[42] In 1955 this policy paid immediate dividends with an increase of Faculty influence on the committee: one of the Faculty's four representatives was henceforth to act as its vice-convenor. Archibald Goodall duly became vice-convenor during 1956.[43] Goodall was also permanent member of the Faculty Council and committed to the re-forming agenda. He now became a linchpin of this reform, keeping the University on side.

As Shaw observed, however, postgraduate education was a long-term strategy for the Faculty and it would be a drain on the slender resources, rather than a profit-making venture, for many years. Finances continued to be a pressing problem. While the Faculty's trust funds amounted to £114,853 at the end of 1955, only the interest from these was available (£3395, mostly from the Lock Funds), and then only under certain limitations. Most of this

---

[41] Craig, *History RCPEd*, p. 765.

[42] See Shaw, 'Memorandum on the Present State of Postgraduate Medical Education and its Organisation in Glasgow', 9 March 1960, Shaw Papers.

[43] *Council Minutes*, 18 January 1955.

money went on improving the library as a postgraduate study facility. The Faculty achieved a surplus of £285 in 1955, but, as the Honorary Treasurer commented, this was 'largely artificial – that in fact the general funds of the Faculty are not sufficient at present to maintain it as a going concern and there is still an obligation on all Fellows to contribute as liberally as possible'.[44] Yet voluntary subscriptions had fallen off after the initial appeal. Therefore Stanley Alstead launched his campaign to introduce compulsory subscriptions at the November 1955 Faculty Meeting (AGM). Although Goodall actually opposed Alstead's motion in Council, he was unsuccessful.[45] The Faculty approved of the principle at the 7 November meeting and in 1956 the Faculty approved amending the regulations to levy a compulsory subscription at the following rates:

Fellows resident in the United Kingdom	£5 5s. 0d.
Fellows resident outwith the United Kingdom; Fellows holding regular Commissions in H. M. Forces; Fellows holding whole-time non-clinical academic or research appointments	£3 3s. 0d.
Fellows, (other than those admitted on basis of their published work), up to a period of five years after admission to the Fellowship; Fellows permanently retired from practice.[46]	Exempt

The compulsory subscription was to take effect from 1 October 1956. It put the Faculty on a firm financial footing and was a key part of the modernisation process.

The year 1956 also witnessed an important new initiative in the Faculty's postgraduate teaching. The 1955 lecture series had attracted very poor attendances, no doubt because now a fee had been introduced, so it was decided to abandon this strategy. The reformers, however, were soon to come up with an even better idea. Goodall and Shaw had been invited to a conference on cardiology in Dundee put on by the Professor of Medicine there, Ian Hill. Hill's conference was the first of its kind in Scotland (and possibly in the whole of the UK). Its main purpose was to enable Scottish practitioners to find out about the latest American advances in cardiology and cardiac surgery. Shaw recently described the impact of this example:

Archie and I were invited as guests and we came home after a couple of days there, our heads spinning. This is what we must do we realised. Leave the money

---

[44] 'RCPSG: Honorary Treasurer's Annual Report, 7 November 1955', in RCPSG, 1/1/4/14.
[45] See *Council Minutes* 18 October and 29 December; and *Faculty Minutes* 7 November 1955.
[46] *Council Minutes*, 20 March 1956.

for the moment. We could provide the organisation and administration, but we did need powerful University help to provide the lectures and the surgical material. Our line was exactly what had been done in Dundee, we decided, curiously enough, to go for the same subject, but to make it much wider. We got Arthur Mackey who was just beginning to start cardiac surgery in Glasgow,[47] and Bert [Barclay].[48] They between them knew a lot of people we could invite as guest speakers and we sent widespread invitations to Universities and Colleges to come to a conference in Glasgow, free. No fee, free meals, and a slap-up dinner at the end of it. We had to turn people down![49]

Goodall took the early lead in organising the conferences. He first suggested the idea to Council, and won their approval to go ahead, on 17 April 1956. He said that didactic lectures were to be discontinued and that 'symposia' should be held instead, funded with the proceeds of the Geoffrey Balmanno Fleming Fund, intended primarily for registrars and consultants. He then advised the University GPGMEC of the plan and convinced them to run the conferences jointly under the aegis of the Committee. This was important as it meant that University staff and facilities in the university units in the teaching hospitals, and elsewhere, could be used. For instance, in the first conference on cardiology, all the leading Glasgow cardiologists spoke or chaired sessions, as well as invited experts. Then, on the last two days, there were operating sessions (at the GRI, GWI, Hairmyres Hospital, Mearnskirk Hospital and the Southern General) and outpatient sessions (at the GRI, Stobhill, the Southern General, the GWI and the Royal Hospital for Sick Children).[50] There was another reason why the reformers were careful to get the approval of the University. As Shaw has recently commented:

They [the University] were quite happy for us to have our own ideas, and, of

[47] One of Illingworth's 'boys' at the University Surgical Unit at the GWI, Mackey was originally Resident under Archibald Young (Regius Professor of Surgery at the GWI from 1924) and rejoined his staff after a period under Muir in Pathology and a Rockefeller Scholarship with Evarts Graham in St Louis. Both before and after the war he held an additional appointment as surgeon to the Southern General (where Shaw knew him). In 1953 he was appointed to the St Mungo Chair of Surgery at the GRI. See MacQueen and Kerr, *The Western Infirmary*, pp. 147–48.

[48] A consultant cardiologist from Mearnskirk Hospital. Dr Shaw commented in January 1999: 'R. S. Barclay began his medical career as a TB Physician and later became a chest surgeon, eventually switching to Cardiac Surgery. He established an excellent unit and team at Mearnskirk, and travelled widely, visiting Cardiac Surgery centres all over the world'. Personal communication from Dr Shaw to Andrew Hull, January 1999.

[49] Interview with Gavin Shaw.

[50] 'University of Glasgow, Royal Faculty of Physicians and Surgeons, Postgraduate Medical Education Committee: Conference on Medical and Surgical Cardiology' (October 1956), programme, Shaw Papers.

course there were many activities that we didn't involve the University in at all, but we always passed them through because they were under the aegis of the Postgraduate Committee of the University of Glasgow and the RFPSG, and we were very glad to see *both* names at the top of the paper![51]

Such cooperation was, however, not difficult to achieve since in this period the key individuals in the Faculty and the University were increasingly the same people: Goodall was vice-convenor of the GPGMEC and Wishart (who had originally arranged the University grant to the Faculty) was postgraduate Dean. The University, preoccupied with undergraduate education, yet needing to keep a supply of trained men for the region, was only too happy to let the Faculty control postgraduate education for it.

The first conference was a tremendous success. While it raised no revenue itself, it was an impressive piece of public relations which loudly announced that Glasgow was a credible centre for advanced postgraduate education. Conferences now became an ongoing feature of the Faculty's postgraduate portfolio. Excepting the Dundee precedent, the Faculty Conference series was the first pioneering step by British medical corporations to become actively involved in broadly-based postgraduate activities which served not only as forums for learning but also for the development of research by the exchange of views. The Edinburgh Physicians, for example, did not begin such symposia until as late as 1961.[52]

In addition to the conferences, the Faculty also supported the new *Scottish Medical Journal* (an amalgamation of the *Edinburgh Medical Journal* and the *Glasgow Medical Journal*, which the RMCSG was unable to continue to support financially). The Faculty contributed £100 to the new journal out of the Lock Funds on Alstead's motion in 1956, and a further £100 in 1957.[53] The new journal meant the reinvigorated Glasgow Medical School had a dedicated research outlet for its new productive academic unit staff. Established access to such a journal has been identified as one of the essential features for maintaining a successful 'research school'.[54] While the new journal had a wide remit and wished to appeal to 'the family doctor no less than to those whose work is highly specialised', half of its twelve man 1956 board of management was drawn from the Glasgow region and included such familiar names as Alstead, Anderson (as chairman), Oastler, and Stevenson.[55]

[51] Interview with Gavin Shaw. Original emphasis.

[52] See Craig, *History RCPEd.*, p. 759.

[53] *Council Minutes*, 18 September 1956, and (from RCPSG, 1/1/4/16), 19 March 1957.

[54] Jack Morrell, 'The Chemist Breeders: The Research Schools of Liebig and Thomson', *Ambix*, 19 (1972), pp. 1–46.

[55] See editorial *Scottish Medical Journal* (*SMJ*), 1, no. 1, January 1956, p. 46, by Thomas Anderson and Rae Gilchrist.

Table 3. *The Conferences, 1956–67*

1	Medical and Surgical Cardiology	1–5 October 1956
2	Gastroenterology	26–30 November 1956
3	Urology	1–5 April 1957
4	Psychiatry	28 October–1 November 1957
5	Neoplasia	26–28 March 1958
6	Haematology	16–20 March 1959
7	Thyroid Gland	11–13 November 1958
8	The Kidney	30 March–1 April 1960
9	Accident Surgery	30 November–2 December 1960
10	Atherosclerosis	8–10 March 1961
11	Diseases of Genetic Aetiology	15–17 November 1961
12	Tissue Transplantation	21–23 November 1962
13	Blood Transfusion	9–10 May 1963
14	Gastroenterology	1–3 December 1965
15	Aspects of Virology	22–23 September 1966
16	Organ Blood Flow	Easter 1967
17	Ischaemic Heart Disease	9–10 November 1967

At the same time, the campaign to reform the government and administration of the Faculty was also progressing. In 1954 the new office of Honorary Secretary was created to deal with the Faculty's developing postgraduate role. The new officer had a place on the Council, and the first incumbent was Dr Alexander H. Imrie. In 1956 attention turned to Council reform. In May Alstead opened the discussion he had called for, at the Faculty meeting on future administration. Thomas Anderson questioned the need for a large Council, suggesting it be slimmed down to six. He also proposed that officers should not automatically be members of Council by virtue of their office (*ex officio*), and that Council committees should be made up of Fellows who were not on Council.[56] This debate continued into early 1957, and at the Faculty's February meeting a substantial measure of reform was won by a majority of thirty votes to seventeen. Previously the Council had consisted of the chief office bearers (the President, Visitor, Honorary Secretary, Treasurer, and Librarian) *ex officio*, plus ten other Councillors. In the interests of

[56] *Faculty Minutes*, 7 May 1956.

administrative efficiency it was now slimmed down to the chief office bearers (no longer *ex officio*) plus just six other Councillors, *all* by election. This meant that there would no longer be a representative of the Licentiates in Medicine and Surgery (TQ), or of the Licentiates in Dental Surgery on the Council. Each Councillor was to retire annually and to be eligible for ree-lection for four years in succession only, following which a year off would have to be taken.[57] These measures were clearly strategies to increase the power of the younger Fellows, and to head off any attempts to thwart the reforming agenda by traditionalists. Shaw indicates that die-hard opposition to change was not very potent, but was there:

> I think a lot of it was old-fashioned conservatism. The organisation had lasted three hundred and fifty years and there was no reason why it couldn't go on ... There was a lot of getting onto the Council just for the status of being on the Council and not really either attending meetings very well or taking much interest beyond that. I don't remember that the opposition was very strongly against us, but they were a bit frightened of what we might do! Indeed for several of the things that it was agreed could go ahead, the then Visitor, Dr Andrew Allison Senior ... was very much for us, but he was put in as our minder. We had to contact him if we were making some proposals. It was very interesting! But we accepted it because we thought in that way that we would carry them with us.[58]

These measures were incorporated into the regulations in March, though the Council did insist that it was necessary to have the GMC representative as an *ex officio* member of Council (as other corporations did), because it was crucial that he transmit current Council viewpoints to the GMC, and vice versa. The Council rejected Alstead's idea of two separate colleges (one for Physicians and one for Surgeons) within the Faculty. This was an incar-nation of Alstead's long-held view that there should be national colleges, and also possibly a reaction to the pressures of specialisation, which had led to the formation of special sub-faculties/committees for specialists in the London Colleges.[59]

Financially 1956 was a satisfactory year for the Faculty. Receipts from candidates had increased by £560, the large majority of which (£367) was from increased take-up of the Primary Surgical Examination. Voluntary contributions were also increased, now yielding £897, and, of course, the

---

[57] *Faculty Minutes*, 4 February 1957, RCPSG, 1/1/4/16.

[58] Interview with Gavin Shaw. Allison was only a 'minder' in relation to the organisation of the conferences, personal communication from Dr Shaw to Andrew Hull, January 1999.

[59] See *Council Minutes*, 19 June and 18 September 1956, and (from RCPSG, 1/1/4/16) 19 March 1957. On the last point see, Stevens, *Medical Practice*, chapter 8.

compulsory subscription came in from 1 October. While the Faculty recorded a modest surplus of £69, again this was only achieved by the continuation of the grant for postgraduate education of £1000 from the University.[60]

Table 4.1. *Results of Fellowship Examinations, 1952–58* [61]

Examined qua Physician	Entered	Passed
1952	35	12
1953	38	14
1954	42	14
1955	31	10
1956	37	14
1958	64	21
Examined qua Surgeon (Primary)		
1952	103	32
1953	164	33
1954	96	32
1955	89	31
1956	154	43
1958	445	107
Examined qua Surgeon (Final)		
1952	4	2
1953	8	4
1954	6	3
1955	16	8
1956	14	9
1958	20	11

[60] Honorary Treasurer's (Dr J. A. W. McCluskie) Report, *Faculty Minutes*, 5 November 1956.
[61] Source: 'RFPS, Table of the Results of Fellowship Examinations', 17 September 1957, RCPSG, 1/1/4/16. For 1958 figures, 'Annual Return to General Medical Council', 11 March 1959, RCPSG, 1/1/4/18.

Table 4.2. *Fellows Admitted, 1947–56*[62]

1947	63
1948	79
1949	65
1950	32
1951	19
1952	13
1953	19
1954	21
1955	17
1956	28
Total	356
	(of which only 27 or 7.6 per cent admitted on basis of published work rather than by examination).

The year 1957 was a key one in the transformation of the Faculty into its modern incarnation as a postgraduate teaching and examining body. Through the efforts of the reformers, the argument for reorientation had been won: the success of the conferences was public proof that the Glasgow Faculty could be a postgraduate centre. The attitude of 'What has Glasgow really got to offer ... No one would really want to come to my wards', that Shaw detected among senior hospital staff in the early 1950s had been overcome.[63] Now was the time to build on these foundations, especially as Alstead had become President in November 1956, with Arthur Jacobs as his Visitor, thus beginning a succession of particularly progressive President/Visitor partnerships.[64] The new role was inextricably linked, as we have noted, with the revival of the Fellowship and hence the revival of the Faculty's finances, and a renewed sense of purpose. With the increasing financial stability of the Faculty after 1956 (compulsory subscriptions and the increasing availability of Trust Fund monies – especially the Lock – for general purposes), and with the stabilisation and growth of the number of entrants for the Fellowship after the initial post-war demand from returning doctors had

[62] Source: 'Memorandum from Dr Shaw to the Council on the Admission of Fellows by "Special Examination" (Regulation Chapter 2 Paragraph 6)', insert with Council Minutes, 15 January 1957, RCPSG, 1/1/4/16.

[63] Shaw, 'Present State of Postgraduate Medical Education and its Organisation in Glasgow'.

[64] Alstead/Jacobs became Jacobs/Wright then Wright became President.

[65] *Faculty Minutes,* 7 January 1957.

subsided (see Tables 4.1 and 4.2), a desire to build a wider postgraduate *school* in Glasgow with an educational function apart from considerations of Faculty welfare can be detected in the debates. We must remember that the generation of doctors now increasingly concerned with the running of the Faculty also had wider loyalties to the University, and to a modern conception of medicine as specialised and scientific (though perhaps based on general medicine and surgery). This was the post-Hetherington generation, when the ultimate goal of many practitioners was a place doing specialised research work in a hospital university unit and within the academic stream of medicine. This kind of career was now possible, and even those whose bent was more towards the clinical care of patients in an NHS post, were still aware that the academic ethos and method now dominated the advance of medical knowledge. Most would therefore now cultivate their own specialty and attempt to conduct some research to advance their own careers, as it was against the academic priorities of teaching and research that suitability for advancement was now increasingly measured. It was, therefore, in the interests of all hospital practitioners that Glasgow should develop as a vibrant centre for postgraduate training and work. This would mean that training was locally available, would mean a good supply of highly-trained staff to the hospitals, and would establish a self-reinforcing research culture, particularly if training was not seen only as for exams but as part of the ongoing medical education of all hospital doctors.

Once again, it was Alstead who made the running in focusing the interests and desires of the younger Fellows, harnessing them to his own, similarly motivated, reform programme. In January 1957, shortly after becoming President, he produced a memorandum for discussion by the whole Faculty on 'The Need for a Younger Fellows Committee to Foster Greater Interest in the Affairs of the RFPSG'.[65] This prompted the younger Fellows to call an independent meeting in Faculty Hall on the 14 January to which sixty junior Fellows came. A committee was formed to coordinate future meetings and, at a further meeting on 18 February, the main calls were for increased Faculty efforts in postgraduate education.[66]

Such a background of pressure from the younger Fellows added a new sense of urgency to the Faculty's consideration of how to further the postgraduate agenda. At the 19 February Council meeting Alstead's memorandum 'On the Need for *Immediate Developments* in Postgraduate Instruction' was discussed and approved.[67] Goodall, as vice-convenor of the (University) GPGMEC, was instructed to convey the President's views to the University

[66] Ibid., 4 February 1957.
[67] *Council Minutes*, 19 February 1957. My emphasis.

and to add the Council's opinion that, 'all aspects of Medicine, Surgery, Obstetrics and Gynaecology should be taken into account when postgraduate courses are being organised'.[68] This amounted to the Faculty telling the University that it wanted postgraduate instruction in Glasgow to be a complete and systematic training for the Glasgow Fellowship, and beyond. At the President's request, the Faculty meeting on the 4–5 March was dedicated to a general discussion of the 'further development and integration' of postgraduate education in Glasgow and the west of Scotland for all practitioners from GPs to registrars in training, through to full consultants continuing their education. This was itself an expanded conception of the scope of postgraduate education desirable in Glasgow, anticipating by forty years the current focus on Continuing Medical Education (CME). Fortunately, an unusually full record of this discussion survives. Alstead himself had a précis made and circulated to all Fellows, 'to promote continued discussion among groups of Fellows in the hospitals and elsewhere'. As he wrote in the introduction to this synopsis:

> Continual interchange of ideas is clearly desirable because it would be difficult to exaggerate the importance of postgraduate education. In short, the enterprise has intrinsic merit – a statement which does not call for amplification in a teaching centre like Glasgow; and secondly, widespread activity in this field would do more than anything else to stimulate the corporate life of the Royal Faculty.[69]

The contributions to this discussion demonstrate the large scope and potential variety of the all-encompassing rubric of postgraduate education. One contributor stressed how keen GPs were to participate, and how this could serve to bring the increasingly divergent career paths of GPs and consultants back together, thus encouraging a new 'wholeness' in the profession. This broader concept of postgraduate education was one which the Faculty was to embrace in the future with the inception of more diploma courses, largely aimed at GPs and foreign students. A start was made in this direction during 1957 with the development of the Diploma in Child Health. This was largely the work of Eric Coleman, Registrar (and later consultant) in Child Health at the Royal Hospital for Sick Children in Professor Stanley Graham's academic unit. The course was largely aimed at and attracted GPs and foreign students.[70]

Two contributions to the above debate in particular, however, capture the contemporary context of the importance of postgraduate education to the

[68] Ibid.
[69] 'The West of Scotland as a Postgraduate Centre: Its Further Development and Integration', Memorandum from the President to the Fellows of the Faculty, 7 March 1957, Shaw Papers.
[70] Telephone conversation between Andrew Hull and Dr Eric N. Coleman, May 1997.

whole Glasgow Medical School. First, Dr Gerald Timbury,[71] a young registrar at the Western Infirmary and a friend of Shaw's, reminded the meeting of the 'realistic needs of the younger members of the profession':

The potential specialist required: (1) additional qualifications, (2) publications – preferably related to work in hand for a thesis (MD, ChM), (3) clinical experience in the main subject and its specialties, and (4) time for study – leave of absence from routine duties in the wards in order to take suitable courses elsewhere. The Royal Faculty should encourage the organisation of coaching classes ('cram courses') to last for say four weeks and designed specifically to obtain a high pass rate in the examinations for higher diplomas. Discussion groups and journal clubs should be set up in all teaching hospitals. There should be a panel of Advisers to help younger men in the design of research projects.[72]

Illingworth added that the next step for postgraduate education in Glasgow should be to organise attachments for postgraduates to clinical units. Illingworth, as we saw, was one of Hetherington's professorial appointments who had done and was doing most to develop the concept of specialised academic medicine in Glasgow. He was keen that postgraduates should absorb the new way of working at an early stage and the university clinical unit was the beacon that transmitted this new ethos to the whole hospital sector. Developing the point of the importance of the units, which was strongly supported by Alstead in his concluding remarks in the memorandum, Illingworth added that once in the unit, 'the postgraduate would then be assimilated into all the activities of the unit on a footing comparable with that which applies to Assistants. The next step would be visits by postgraduate students to other units so that they might avail themselves of special facilities – demonstrations, discussions, operating sessions …'[73]

Clinical attachments for postgraduates were usually meant for GPs and foreign students, since, at least in principle, those wishing to be consultants had secured training grade positions on the NHS career ladder (as Senior House Officers and Junior and Senior Registrars, or Junior and Senior Hospital Medical Officers). There were, however, many more doctors wishing to become consultants than there were places on the training ladder. Clinical

---

[71] Timbury later became a consultant psychiatrist and supervised the referral service for out-patients and a consultative service for in-patients at the GWI (though there were no psychiatric beds there). In 1965 Physician Superintendent at Gartnavel Royal (Mental) Hospital, he oversaw the development of Gartnavel as a specialist psychiatric institution for the Western Region with the fusion of the Gartnavel and GWI BOMs in 1968. He later succeeded Cruickshank as Postgraduate Dean. See MacQueen and Kerr, *The Western Infirmary*, p. 140.

[72] President's Memorandum, 'The West of Scotland as a Postgraduate Centre'.

[73] Ibid.

assistantships were also, therefore, an informal way for those who were forced for economic reasons to be GPs, but still entertained hopes of consultant status, to assemble a portfolio of varied clinical experience at a specialist level. This experience might lead to consideration for a junior hospital job; it would in any case further the GPs continuing medical education.

Alstead entertained (ultimately unsuccessful) plans to formalise the status of the clinical assistant as intermediate grade between the GP and the consultant, realising that, because of the number of young doctors, some formalised mechanism for translation between the two career paths was necessary. This formalised clinical assistantship was also seen by Alstead as an alternative to the SHMO grade, which had been introduced for older GP specialists and those trainee hospital doctors who had failed to secure the all-important Senior Registrar grade, but which had rapidly fallen into disrepute as a sub-consultant cul-de-sac whence too many juniors chased too few jobs. Alstead's idea of a redefinition of the clinical assistantship was echoed by the Platt Committee which sat from 1958 and reported in 1961. It recommended a second grade below consultant – the medical assistant. This grade would perform the same duties as the discredited SHMO but would be designed specifically to provide general practitioners with access to the training ladder, preparing them for translating to consultant posts.[74] What is less clear is whether Alstead and the Faculty envisaged that those who already held hospital posts would obtain study leave to take up a number of short clinical assistantships to increase their experience of the various specialties and units.

The debate in Faculty inspired a succession of memoranda on the question of how best to organise a comprehensive postgraduate course. Stress was laid on the importance of the Faculty acting to encourage contacts between young researchers and chiefs, thereby stimulating an ongoing research culture in Glasgow.[75] But the main thrust of these memoranda was that Glasgow's reputation as a postgraduate medical school should be based, as Illingworth had earlier suggested, on attachments to clinical units.[76] Partly this was because the Edinburgh Postgraduate Board already ran a systematic course in internal medicine, which was constantly oversubscribed, and it was pointless to compete with this.[77] Also, the higher qualifications which the

[74] See Stevens, *Medical Practice*, pp. 150–52.

[75] 'Postgraduate Education in Glasgow', a memorandum by T. J. Thomson, March 1957, Shaw Papers.

[76] See the above memorandum and Shaw's, 'Postgraduate Medical Education in Glasgow: The Next Step', 25 March 1957, Shaw Papers, which was sent to all Council members.

[77] This is directly mentioned in Thomson's memorandum. See also Craig, *History RCPEd*, p. 765.

students would be sitting all included a formal examination at the bedside in clinical medicine which had to be passed independently of the other parts of the examination. There were also more deep-seated reasons for the emphasis on these clinical attachments. In the same way that Hetherington's appointees had continued to stress the importance of clinical experience, partly out of deference to the local clinical elite's medical ideology, so the proposed scheme of clinical attachments was also founded on the central importance of clinical work both as an educational model and as the continuing basis of medical knowledge. However, since the rotating attachments would be held in the clinical units, most of which were, increasingly, subject to academic influence, students would absorb the importance of clinical work against the background of the specialised and the scientific. Rotating attachments would also impart the sense of general medicine as the core. The conception of clinical work which trainees gained would thus be an eminently modern one, but one rooted in the traditional strengths and orientations of Glasgow medicine.

After the debate, Goodall informed Wishart at the GPGMEC so that a continued joint front was maintained. Advertisements for clinical assistantships had appeared in the *BMJ* and *Lancet* by May. The WRHB was also to be informed of developments and tested on the possibility of providing study leave for juniors.[78]

Efforts were also made during 1957 to continue the public relations campaign for the Fellowship. As Goodall noted, 'the most important action in this field is the establishment of a worthy postgraduate school which will attract students to Glasgow and which will encourage them to take the Faculty examinations'.[79] While this was being established there were more immediate steps which could be taken. Problems of recognition of the Fellowship as a passport to consultant status were still being experienced, both in the UK (the Liverpool Regional Hospital Board was still uncooperative) and abroad (the Faculty had only just obtained recognition from the medical authorities in Pakistan, and was keen to secure the same in India and Ceylon). A statement describing the status and examinations of the Faculty was therefore drawn up to be circulated to all Scottish Regional Hospital Boards, to government departments, and especially to the Colonial Office. The statement stressed the equivalence of the Faculty's examinations with those of London and Edinburgh.[80] During 1957 there were also reviews of examination procedure

[78] *Council Minutes*, 16 April and 21 May 1957.

[79] Minutes of a Meeting of the Committee on the Status of the Fellowship, 11 April 1957, RCPSG, 1/1/4/16.

[80] See memorandum entitled simply, 'Royal Faculty of Physicians and Surgeons', inserted with the 11 April minutes of the above committee.

in the Surgical Fellowship in special subjects, and in the admission of Fellows on the basis of their published work, under chapter two, paragraph six of the regulations.[81]

The year ended with Gavin Shaw becoming Honorary Secretary in November. The younger, more progressive element was now firmly in control of the Council: Alstead was President, Jacobs Visitor, Shaw Secretary and Goodall permanent Council member and deputy-convenor of the GPGMEC. Alstead praised the contribution made by Fellows under forty in his 'Letter to Fellows'. They were the new 'source of strength' to the Faculty. He further warned older, more traditional elements that:

> Tradition is comforting and encouraging and it is not to be despised. But tradition should not be permitted to cloud our appreciation of responsibilities facing us in 1957: it would be a dangerous thing if it were to inhibit realistic planning for the years immediately ahead.[82]

The policy that these reformers had been pursuing was also now beginning to reap financial rewards. Fellowship income was dramatically up in 1957: the Medical Fellowship by £284 and the Surgical Fellowship by £1197. That other modernising innovation, annual subscriptions, was now bringing in £1926, and the University had confirmed that it would guarantee the annual grant of £1000 for postgraduate education for a further five years till 1962.[83] No wonder then that Alstead spoke of 'signs of vigorous life' in postgraduate education, and proclaimed the now familiar message that 'the prestige of any medical centre depends not only on the status of its undergraduate school, but also on the provision it makes for those seeking higher education in the specialties'.[84] In other words, not only the University, but now also the Faculty (as postgraduate teaching and examining centre) was once again crucial to the health of the Glasgow Medical School.

One thing still lacking, however, as Jacobs had pointed out at the June Council, meeting was a modern lecture theatre of 150–200 seats in which to hold the expanding postgraduate activities.[85] An architect (W. N. W. Ramsay) had been engaged to prepare plans for a theatre to be built on the vacant land behind the East House (No. 236) in St Vincent Street, but this was all very

---

[81] See *Council Minutes*, 15 January and 21 May 1957.

[82] Stanley Alstead, 'President's Letter to Fellows', *Amalgamated Minutes*, 30 October 1957, RCPSG, 1/1/4/16.

[83] *Faculty Minutes*, AGM, 4 November 1957, Honorary Treasurer's Report.

[84] Alstead, 'President's Letter to Fellows'.

[85] *Council Minutes*, 18 June 1957.

exploratory without the considerable funds needed for the venture being available.[86]

This need was met in February 1958 when Alstead was able to report that Sir Maurice Bloch had offered the enormous sum of £20,000 to pay for a lecture theatre.[87] It appears that Jacobs was the prime mover in securing this welcome gift. It was clearly on his mind at the end of the previous year, and he was the chairman of the Lecture Room Committee which was trying to find ways to materialise the plans. Bloch, the Honorary President of the Glasgow Jewish Board of Guardians, had made his money in the whisky trade. He had already, through the Sir Maurice Bloch Trust (which had funds of over £400,000), given £35,000 to the University to establish a Fellowship in Virology with special reference to poliomyelitis.[88] It seems that Jacobs approached Bloch and persuaded him that investment in the Faculty would fulfil an educational function central to the future of the whole Glasgow School, and would therefore also be a positive piece of public relations. The contractor (Morrison Dunbar) began work on the theatre in October 1958. Also in October Andrew Allison, the convenor of the Property Committee began to implement a five-year plan which he had drawn up for the complete refurbishment and redecoration of the Faculty premises, to bring them up to the standard to be set by the new theatre.[89] The modernisation of the Faculty's role was therefore matched by an overhaul of its administrative machinery and also of the very fabric of the buildings. The Maurice Bloch Lecture Theatre was completed in 1959 and opened by the donor (who had been created an Honorary Fellow for his services to the Faculty) on 29 September 1959. Alstead was present at the ceremony and gave an address thanking Bloch. Also tellingly present was Sir Hector Hetherington, whose comprehensive policy for rejuvenating Glasgow medicine had made possible the Faculty's reorientation as a postgraduate teaching and examining body.[90] As Gavin Shaw has recently commented, after the construction of the Maurice Bloch Lecture Theatre 'We now had the groundwork of an educational policy ready for us'.[91]

Meanwhile the plans for a comprehensive scheme of postgraduate education were also progressing. Alstead's radical ideas on the function and status

[86] RFPSG, Committee to Consider Lecture Room Accommodation, Minutes 14 November 1957, RCPSG, 1/11/5/4.

[87] *Council Minutes* 18 March 1958, RCPSG, 1/1/4/17.

[88] 'Announcement for Press Association', RCPSG, 1/11/5/4. The gift was recorded in the *Glasgow Herald*, 22 February 1958.

[89] For details see Hutchison, *Property and Environment*, especially pp. 124–31.

[90] Programme for 'Opening of the Maurice Bloch Lecture Theatre', RCPSG, 1/11/5/4.

[91] Interview with Gavin Shaw.

of clinical assistantships seemed to have proved difficult to implement. They remained largely for foreign students who had no other access to hospital appointments. However, there was still a strong feeling in the Faculty that a meaningful clinical component to the postgraduate programme for those wishing to sit higher qualifications should be devised. The Honorary Secretary, Gavin Shaw, suggested to Council that 'Ward Round' Courses should be established.[92] The chiefs or Senior Registrars in the Glasgow Teaching Hospitals (which now included such hospitals as the Southern General) would conduct teaching ward rounds in both medicine and surgery with small groups of ten. There would be three sessions of ten rounds, one in each week of the university teaching terms, and students would pay the teachers a fee of three guineas for each one and a half hour period of instruction.[93] On Shaw's suggestion, Faculty Council recommended to the GPGMEC that a committee of the younger Fellows be appointed to organise these courses. The GPGMEC approved and Shaw became convenor of the committee. His fellow members were Goodall, Robert Fife, William Sillar, Douglas Clark, Thomas Semple and Thomas Thomson. This Committee fulfilled Shaw's hopes that a permanent Faculty standing committee of young Fellows (rather than Council members) could be established to run the Faculty's involvement in postgraduate education.[94]

The courses, which began in October 1958, were very successful in medicine but the surgical course did not attract enough students and had to be discontinued. The first course proved very expensive to run with the low fee of three guineas. The Faculty sustained a loss of £287 on the first year. Unfortunately, due to the death of Wishart, the Postgraduate Dean, it was not made up by the GPGMEC as had been agreed. Nevertheless, the course ran during 1959–60, though with the fees necessary increased to five guineas and yielded a small profit of £125. Like the rest of the Glasgow postgraduate programme, the courses were entirely organised by the Faculty, merely approved by the GPGMEC.[95] A surviving programme from 1960 shows that the GRI, GWI, Southern General, Victoria Infirmary, Stobhill, Mearnskirk, Ruchill, and the Royal Hospital for Sick Children (RHSC) were all involved.[96]

[92] Shaw, 'Present State of Postgraduate Medical Education'.

[93] Minutes of Committee Appointed by Council to Organise Postgraduate Teaching Rounds, 28 April 1958, Shaw Papers.

[94] See Shaw, 'Present State of Postgraduate Medical Education'.

[95] Ibid.

[96] The teachers in 1960 were: in cardiology, J. H. Wright at the GRI and Professor E. J. Wayne and Olav Kerr at the GWI; in haematology, Professor L. J. Davis and A. Brown at the GRI and McCluskie at the GWI; in general medicine, Andrew Allison (junior) and Scott, Shaw and Lazarus at the Southern General, Albert A. F. Peel at the VI, Professor Alstead, Rogen

The Ward Round Courses continued until 1973 when low attendance forced their curtailment.[97] By this time the introduction of rotational training schemes had ensured broad clinical experience for all those training to be consultants of all types.

The Faculty's financial position during 1958 was much improved, largely due to increased numbers for the Fellowship (see Tables 4.1 and 4.2). Moreover, Alstead had introduced a further new source of funding: he had begun to approach commercial firms for donations specifically to fund postgraduate education. By 16 September he had raised £2845 from this source.[98] In 1959 the Faculty received a donation of £1000 from ICI and an interesting offer from Sandoz Products Limited to fund future research projects of mutual interest.[99]

The Faculty's postgraduate programme was completed during 1959 with the introduction of the 'Recent Advances' courses in medicine. These were again devised and implemented by Shaw and the group of younger Fellows which had formed the Ward Rounds Committee.[100] The 'Recent Advances' courses were aimed at Glasgow Senior House Officers and Registrars in Medicine who needed further guidance in their training and who could never get more than two weeks study leave. They were also aimed at foreign students who already had a clinical assistantship but who wished, 'in addition, a course of clinical medicine which would give them the opportunity of learning something from *all* the Glasgow clinical centres in a short time'. Most of these foreign students went on to take the Edinburgh Physicians' ten week intensive course in internal medicine and thus did not require further theoretical, or as Shaw put it, 'full-dress instruction' courses, but did need a more clinically orientated course. Shaw sketched the proposed course thus:

> Lasting eleven or twelve days, it would comprise lecturettes on the most topical of recent advances, clinical demonstrations in the City's hospitals and panel

and J. B. Rennie at Stobhill, Professor Davis and Imrie at the GRI and T. N. Fraser at the GWI; in endocrinology, Ian Murray at the VI, Eric Oastler at the GRI and Wayne at the GWI; in diseases of the chest, R. J. Cuthbert and G. H. Stevenson at Mearnskirk and Andrew W. Lees at Ruchill; in paediatrics, Professor Stanley Graham and J. H. Hutchison at the RHSC, and Ian D. Riley at Stobhill; and in neurology, J. B. Gaylor at the GWI and Gervais J. Dixon at the Southern General. See 'University of Glasgow and Royal Faculty of Physicians and Surgeons Postgraduate Medical Education Committee: Postgraduate Teaching Ward Rounds, January to March, 1960', Shaw Papers.

[97] *AGM Billet*, 5 November 1973, 'Honorary Secretary's Report'.

[98] *Council Minutes*, 16 September 1958.

[99] *Council Minutes*, 17 February 1959, RCPSG, 1/1/4/18.

[100] Details here from Shaw (and the other members of the Ward Rounds Committee, T. J. Thomson, A. Melrose, W. G. Manderson, J. Crooks, and Robert Fife), to Goodall (as vice-convenor of the GPGMEC), 16 March 1959, Shaw Papers.

discussions and clinico-pathological conferences. The course would be designed to stress its clinical nature ...[101]

The morning mini-lectures were coordinated with the afternoon clinical instruction. For example on the first course from 14–25 September 1959, the first morning consisted of lectures on recent developments in aetiology, diagnosis and special techniques in congenital heart disease, and on the treatment of hypertension. The afternoon was taken up with clinical demonstrations at the Royal Infirmary of three cases of heart disease, and the surgical treatment of congenital heart disease.[102] This series of courses continued until 1974, when, because of poor demand, it was replaced by a half-day release course.[103] Once again, the introduction of organised rotation schemes for SHOs and Registrars meant that sufficient experience of modern clinical developments was gained by trainees.

The completion of a comprehensive postgraduate programme was marked at the end of 1959 with the production by the GPGMEC of a postgraduate brochure as a publicity device to attract even more students to the new Glasgow postgraduate school. The brochure was suitably adorned with a picture of the University on the front and one of the Faculty premises on the back, these positions reflecting relative status, though not relative input into the postgraduate programme. Shaw, however, still wasn't entirely happy with the pace of developments. His copy of the brochure is annotated, 'Produced in 1959 *two years* after I had written to the PG Committee representative about it and even sent a fully "mocked-up" copy'![104] The Faculty's success at reinventing itself was also more publicly expressed in Princess Alexandra's acceptance of the Honorary Fellowship. The Princess agreed to accept the Fellowship in May 1959 and it was conferred at a ceremony on 1 June 1960.[105]

Shaw had realised the need for a new clinical course for postgraduates intending to take higher Fellowship or Membership qualifications when the Ward Round courses had not attracted this type of student. The Ward Round courses had really been aimed at trainee consultants from the district general hospitals who felt isolated from the academic ethos and teaching methods of the central teaching hospitals.[106] The courses were a mechanism for bridging the perceived widening gap between the central teaching hospitals and

[101] Short and long quotations above from ibid.

[102] 'University of Glasgow and RFPSG Postgraduate Education Committee "Recent Advances in Clinical Medicine"', 'Outline of Course', 1 September 1959, Shaw Papers.

[103] *AGM Billet*, 4 November 1974.

[104] University of Glasgow/RFPSG, 'Postgraduate Medical Education, 1959', Shaw Papers.

[105] *Council Minutes*, 19 May 1959.

[106] See Shaw, 'RCPSG', p. 105.

the district generals. However, the courses had instead attracted trainees from recently emerging specialties (for example anaesthetics, psychiatry and laboratory services) who needed further practical clinical experience, 'of general medicine as well as a specialty' to sit higher examinations.[107] Numbers of these types of specialists had risen dramatically with the inception of the NHS and the rise of the district general hospitals to teaching status.[108] It was difficult, however, for trainees in these specialties to gain clinical experience at the central teaching hospitals, as advertised posts there would be filled by specialists whose subject was closer to the research interests of these hospitals, and higher up the new hierarchy of the specialties (for instance, cardiology and urology). While the Ward Round courses were to continue for this type of postgraduate (since there was an obvious demand), this was not the end of the Faculty's involvement with such practitioners.

The year 1959 ended with intense debate between the permanent Joint Standing Committee of the three Scottish medical corporations and representatives of the disciplines of pathology, bacteriology, psychological medicine, radiology/radiotherapy, anaesthetics, public health and child health, as to what would be the most suitable type of higher examination to admit to consultant status in their respective areas. At this time there were no recognised qualifications acting as passports to consultant status in these fields, just as there was no definition of a consultant in the older specialties. The increasing specialisation of hospital medicine, which was enshrined in the staffing structure of the NHS, had necessitated a new working conception of medical education for the hospital practitioner in which *postgraduate* education was now the most important stage. Before the entrenchment of the specialisation of medical knowledge in hospital practice the *generalist* higher qualifications of the medical corporations had been a mark of the fully trained physician or surgeon. But increasingly *further* clinical experience of specialisms was necessary. This was obtained by a series of appointments in training grade positions, *after* hospital doctors had taken their higher qualification. The role of the higher qualification thus changed to indicate not the completion of training, but rather the possession of a *potential* to

[107] Shaw, 'Memorandum on a Possible Postgraduate Course in Medicine for 1959–60', February 1959, Shaw Papers.

[108] See 'RFPSG Memorandum on Medical Staffing in the Hospital Service', May 1959, RCPSG, 1/1/4/18. This was the Faculty's contribution to the Ministry of Health Working Party on Hospital Staffing. Chaired by Sir Robert Platt, President of the RCPL, this was made up of eleven representatives of the English and Scottish health departments, the JCC, the three Scottish Medical Corporations and the Central Consultants and Specialists Committee of the BMA. The Platt Report, *Medical Staffing Structure in the Hospital Service*, was published in 1961. See Stevens, *Medical Practice*, pp. 150–52.

acheive consultant status *after* further training. But there was no formal structure to this further training, just as there was no formal defintion of consultant status in any branch of medicine. By the late 1950s this was becoming an increasingly pressing problem. For the medical corporations there was the additional aspect of how their traditional, generalist higher qualifications would synchronise with the training of an increasing variety of specialist groups.

Most of the trainees in the newly-emerging specialisms currently took diplomas from one of the medical corporations or universities. In fact, this whole debate had originally surfaced when the Glasgow Faculty was considering establishing new diplomas in psychological medicine, clinical pathology and laboratory medicine, and had canvassed the opinion of the other corporations.[109] However, diplomas were not perceived to be of a high enough standard, while the higher qualifications of the ancient medical coporations were largely generalist. Unlike other specialties (such as orthopaedics, neurosurgery, plastic surgery and urology), these newly-emerging specialties were not yet 'status specialities'. They had no strong commitment to general medical and surgical training, as well as specialised training. They therefore lacked influence within the long-established medical corporations and were constantly tempted to form thier own. Furthermore, there were fewer consultants in the more established specialties than in the newly-emerging specialties which had greatly expanded with the NHS demand. The newly-emerging specialties needed to establish separate training to improve their status and to form a clear career path.

In England these problems had resulted in the fragmentation of both the Colleges. The Surgeons had bowed to pressure from the traditional specialties and had instituted an FRCS in ophthalmology and otolaryngology, and a Fellowship in dental surgery (FDS, RCS) in 1947.[110] The dentists were treated differently in that they got a semi-autonomous faculty within the college. This faculty controlled the postgraduate examinations. A similar faculty and examination were created for anaesthetists in 1948. These specialists continued to take the general primary examination in the basic medical sciences, but were no longer required to take general surgery in the final examination.

The Physicians had experienced similar pressure from the specialties of pathology, paediatrics and psychiatry. However, specialty formation in medicine followed a different path to that in surgery. Medical specialties had a much clearer continuing relationship to general medicine, and thus the

[109] See *Council Minutes*, October-December 1959, passim.
[110] See Stevens, *Medical Practice*, pp. 111–13.

pressures here were less intense. However, obstetricians and radiologists had already broken away to form their own colleges.

In the late 1950s these problems of specialisation and sub-specialisation came to a head in Scotland, with demands for Scottish specialist qualifications of the same standard as the Fellowship/Membership. The problems in the newly-emerging specialties posed wider questions about the conception of the unity of general medicine embodied in the higher qualifications of the Scottish corporations. Were the higher qualifications of the corporations going to continue to be qualifications in general medicine and surgery, to be taken early in the career as an estimation of general ability, *after* which would-be consultants would gain specialist experience and, possibly, qualifications ? This approach risked the fragmentation of the corporations and the formation of individual specialty colleges which might ultimately usurp the role of the ancient generalist corporations. Or would the corporations abandon the commitment to general medicine and surgery, that most of their members had already effectively abandoned in their hospital work, and offer a range of specialist higher qualifications taken later and later in the hospital career as clinical experience in the specialty was built up? One suggestion put forward in Scotland was that the Scottish medical corporations should form an joint umbrella association to which faculties of specialists would be attached, and which would offer joint Scottish specialist Fellowships.[111]

This was a debate not only about the survival of the corporations but about the very nature of medicine itself. It was the also the logical outcome of the founding of the NHS on the specialist consultant working in the hospital. So, what had given the Faculty a vibrant new role now threatened to leave it with no role at all.

[111] C. W. A. Falconer (Secretary of the Joint Standing Committee of the Three Royal Scottish Medical Corporations) to Arthur Jacobs, President RFPSG, 23 November 1959, in RCPSG, 1/1/4/18.

5

The Royal College, 1960–1972

The rapid expansion of medical science which has taken place during the preceding decades and will certainly continue, together with the rising standards of medical care expected by the public and demanded by the profession, have made the further education and planned training of graduate doctors a problem requiring urgent consideration. *It has become obvious that both education and training must be placed upon a more organised footing than was thought necessary in the past.*[1]

[The name change and the MRCP (UK)] brought the College in Glasgow fully into the stream of the other two Medical Colleges ... one became equal in status, perhaps not in influence in terms of size relative to the other two, but certainly it brought the College into that central stream. I don't imagine, looking back to the 1940s that the President of this College would have regular meetings with the London College of Physicians' President, or that he would get on the night train and come to a meeting in Glasgow ... There is now a circuit and the meetings are held in rotation.[2]

The 1960s, even more than the 1950s, were a decade which was self-consciously 'modern'. Rising real wages led to increased demand for new consumer products many of which were highly technological, like the television. Science was fashionable, a buzz-word. A telling indication of the contemporary cultural resonance of science came in the 1964 General Election campaign when the leader of the Labour Party, Harold Wilson, famously tried to gloss his political programme with the importance of science and technology to economic development. At the Labour Party Conference on 1 October 1963, Wilson argued that:

We are redefining and we are restating our socialism in terms of the scientific

---

[1] Scottish Post-Graduate Medical Association, *Report of a Working Party on Postgraduate Medical Education within the National Health Service in Scotland*, 7 April 1967, p. 8, Shaw Papers. The Working Party was chaired by Dr J. H. Wright. 'Education' was taken to mean systematic courses, and 'training' planned accumulation of practical experience in a vocational field in NHS/University hospital posts (or general practice). Original emphasis.

[2] Interview by Andrew Hull with Professor Arthur Kennedy, RCPSG, 30 September 1998.

revolution ... The Britain that is going to be forged in the white heat of this revolution will be no place for restrictive practices or outdated methods on either side of industry.[3]

It is perhaps even more telling that he was widely quoted (and remembered) as having coined the phrase 'the white heat of technological revolution'.[4] But science was not only seen as the answer to Britain's industrial decline. Scientific, specialised medicine was viewed, by both the profession and the public, as the future of medical practice. In a period of optimism and expansion in the NHS, when many new hospitals were planned and some built,[5] and when the number of consultants was rapidly increasing,[6] specialisation was seen, increasingly, not only as an inevitable symptom of the advance of medical knowledge but as the very motor which drove that advance. Medical doctors were also scientists, working in increasingly technological ways on and in the human body. Medicine also attracted much public interest. This was both reflected and fostered by the BBC series 'Your Life in their Hands'. This series, screened from 1958, presented the public for the first time with the marvels of live operations and portrayed technocratic hospital doctors as the masters of high technology medicine.[7] As Marc Berg has recently written, in this period the American medical community

---

[3] Cited from Tom Wilkie, *British Science and Politics since 1945* (Oxford, 1991), p. 39 n. 1.

[4] This is how the Wilson quotation appears in the otherwise very useful Arthur Marwick, *British Society since 1945* (Harmondsworth, 1984; originally 1982), p. 115. Other useful background on this period can be found in Robert Hewison, *Culture and Consensus: England, Art and Politics since 1940* (London, 1995).

[5] Hospital plans for England, Wales and Scotland were unveiled by Enoch Powell as Minister of Health on 23 January 1962. The Scottish plan proposed the building of ten district general hospitals in large population centres, the rebuilding of the teaching hospitals and smaller hospitals to provide acute services for remote areas. Like the English plan, it was the subject of much Treasury-sponsored curtailment by 1964. See Charles Webster *The Health Services since the War*, ii, *Government and Health Care: The National Health Service, 1958–1979* (London, 1996), pp. 92–109. See also Department of Health for Scotland, *Hospital Plan for Scotland*, Cmnd 1602 (Edinburgh, 1962); and John Kinnaird, 'The Hospitals', pp. 213–75, in Gordon McLachlan (ed.), *Improving the Common Weal: Aspects of Scottish Health Services, 1900–1984* (Edinburgh, 1987), especially pp. 254–59.

[6] In the Western Region there were 459 consultants in 1956. This had risen to 875 whole time or equivalent by 1979. In a period when the region's population declined slightly (from 2,909,000 to 2,573,648), this meant that the ration of consultants per 100,000 rose from 15.8 to 34. See Webster *The National Health Service*, ii, appendix 3.24, p. 823.

[7] The first series was transmitted during 1958 and further series followed in the 1960s. The focus was increasingly exclusively on high-technology hospital medicine. I am grateful to Kelly Loughlin of the London School of Hygiene and Tropical Medicine for help on this point.

was positive about the application of science to medical practice, in fact science was seen as permeating practice:

> Editorials from the 1950s and 1960s resonated with an awareness that medicine was living through its 'Golden Years' ... research efforts were exploding in size and number, and the medical profession's ability to frame debates on the future of health care was stronger than ever. The intertwining of professional power and scientific development seemed to hold the promise of indefinite accomplishment and strength.[8]

In Britain too, hospitals were increasingly divided up into an ever-growing number of specialist units. Hospital doctors now needed increasingly to know the application of basic sciences to the problems both of diagnosis and treatment. They also now increasingly rarely worked alone but as part of teams, each participant bringing to the enterprise his or her own special knowledge. Sir Charles Illingworth captured the contrast between surgical practice when he first trained in the early 1920s compared to modern surgery, in a beautifully evocative passage from his autobiography:

> A surgical operation of those days was an event of high drama, appropriately enacted in the arena of a theatre. There, at the centre of the stage, was the concentration of interest, focused on the individual feat of expertise in which the maestro brought his genius to bear on his protégé, the patient. There, at the converging focus of the light beams, stood the surgeon, dominant and incisive; and below him the patient, inert and submissive; around, and in the tiers of seats in the gallery above, the student watchers like a Greek chorus, ready to applaud or condemn. It is very different now. The operation is no longer the work of one man but of a large team. For a major procedure the floor is crowded with assistants, each intent on his own particular duty. The anaesthetist ... manipulates machinery to control the blood volume, the oxygen content, the acid-base balance. Technologists, enmeshed in tubes and wires behind a formidable array of pumps and monitors, check the heart beat and the blood pressure. Behind and around are the engineering devices, the air purifiers, the vacuum pumps and diathermy machines and all the rest of the scientific impedimenta essential for this surgical tour-de-force. And somewhere, lost in the crowd, is the surgeon himself, the creator of this little world of science and now only one of the many actors on the stage, and indeed the least scientific of them all.[9]

Hospital doctors intending to proceed up the Spens career ladder to consultant grade,[10] who would previously have been generalists who cultivated

---

[8] Marc Berg, *Rationalizing Medical Work* (Cambridge, Massachusetts, 1997), p. 19.

[9] Sir Charles Illingworth, *All Men's Lives*, pp. 47–48.

[10] Ministry of Health, *Report of the Inter-Departmental Committee on the Remuneration of Consultants and Specialists*, Cmnd 7420 (London, 1948), the Spens Report on Consultants and Specialists.

a special interest almost as a hobby, now had to develop a serious specialist bent to their education, training, and clinical practice. As Arthur Kennedy recently noted:

> Throughout the 1920s and 1930s physicians might have a particular name that they were good at hearts or stomachs or whatever, but they also considered themselves to be pretty competent all round, and that they could play in any position in the team. It became increasingly apparent after the War, in the 1950s, that what was going to happen was that people would become trained as generalists and then specialise, as a much more positive thing. They wouldn't be simply called a physician, they would be called a cardiologist or a rheumatologist or a nephrologist.[11]

This was demanded not only in academic departments but increasingly also in NHS service work. It was simply not possible anymore for the generalist to be competent in everything; there was too much knowledge and it was increasingly compartmentalised. Such developments in medical knowledge and in the organisation of medical work in the hospitals also meant, however, that specialist training was increasingly difficult to obtain in a series of (so-called) training grade appointments in a succession of different types of unit, in which, in most cases and certainly in the more junior positions, most of the time was taken up with NHS service work and study-leave was a rarity. In 1970 R. B. Wright,[12] then the President of the Glasgow College, neatly summed up this new situation which demanded the formal organisation of training:

> Until the middle of this century the training of physician and surgeons was achieved by a combination of self-discipline and apprenticeship to an acknowledged master of cult or craft. The scientific and technological explosion of the past quarter of a century has resulted in enormous and rapid expansion of knowledge. Specialisation has become narrower and narrower. The general physician and general surgeon no longer profess to encompass the whole range of medicine or of surgery. Increasing dependence on ancillary diagnostic and therapeutic skills has made it necessary for clinicians to work in teams. Such teams

---

[11] Interview by Andrew Hull with Professor Arthur Kennedy, RCPSG, 30 September 1998.

[12] Sir Robert Brash Wright (1915–1981): graduated from Glasgow University BSc (1934), MB, ChB with Honours (1937) and ChM (1953). He served with distinction in the RAMC during the Second World War and was an Assistant Surgeon in the GWI after the war, until he was appointed Surgeon-in-Charge at the Southern General in 1953, a post which he held until his retirement from the NHS in 1980. He obtained the Fellowship of the RCSEd in 1947, and became a Fellow *qua* Surgeon of the Glasgow Faculty (FRFPS Glas.) in 1948, and FRCS Glas. in 1962. He was knighted in 1976. He was a member of the GMC from 1970, and was president in 1980. He published papers on gastroenterology, vascular disorders and postgraduate education. He was President of the Glasgow College from 1968–70. Details from Gibson, *RCPSG*.

tend to concentrate on clearly defined limited therapeutic problems, to attract appropriate clinical cases and no longer see or manage a broad spectrum of medicine or surgery. They can no longer provide a wide basic training for junior staff or they do so on limited material. If graduate training is not to suffer it must be so arranged that the trainee can work in several teams in sequence. He must have opportunity to familiarise himself with many varied techniques and with their associated technology and apparatus. Such training demands organisation by knowledgeable seniors and it can no longer be left to chance or to the initiative of the individuals ...[13]

A further problem for the royal medical corporations, whose generalist higher qualifications had historically served as a passport to higher hospital appointments, was how to accommodate the needs of emerging specialists. After 1948 the profession realised that hospital doctors needed two types of postgraduate education to take up more senior appointments in the new Health Service, especially those in which they would begin to exercise some degree of independent clinical judgement. They needed cumulative clinical experience in a succession of posts of increasing seniority on the new Spens training ladder, and they needed systematic education through lectures, seminars and clinico-pathological conferences to broaden and deepen their understanding of a rapidly advancing medicine. The latter had begun to be organised in some areas, and in Glasgow by the Faculty and the University under the auspices of the Glasgow Postgraduate Board. However, though Glasgow (and Edinburgh) was advanced in postgraduate provision and had a (largely) harmonious administrative set-up which facilitated the ironing out of logistical problems, elsewhere things were not so good. The establishment of new medical schools in England (at Leicester, Leeds and Nottingham for example) made the problem more pressing. Postgraduate education was loosely governed by the same *laissez-faire* principles which had applied to undergraduate medical education before 1858. The related factors of the existence of the NHS, advances in medical knowledge and changes in medical practice made a similar overhaul of self-regulation necessary. As in 1858, this was not begun by the state, but by the profession itself, which was mulling over such issues and the best solutions throughout the 1960s. At the same time, specialisation was reducing the corpus of general medicine and surgery to a dwindling rump. How could the medical corporations' higher qualifications remain generalist? Indeed what was *general* medicine and surgery anyway, and what was the relationship of this knowledge, however defined, to what the specialist needed to know, assuming that

---

[13] R. B. Wright, 'The Organisation of Postgraduate Medical Training', *Journal of the Royal College of Surgeons in Ireland*, 6 (July 1970).

that question could be first answered. If the higher qualifications remained generalist, even with an expanded definition of this concept, there was a strong possibility that new specialist groups, hungry for professional identity and status, would break away to establish their own college and examination system. This had already happened in 1929, with the Obstetricians and Gynaecologists, and was about to happen with the Pathologists (1962) and Psychiatrists (1971). As one consultant wrote in 1963: 'It seems quite unacceptable that keen and aspiring ophthalmologists should be failed because they do not know the minute anatomy of the pudendal nerve, the psoas major, the ext. pollicis longus, and the pubic bone'.[14]

On the other hand, abandoning the traditional attachment to generalism would involve a break with the broadly unified and holistic concept of medicine that the royal medical corporations, as the institutional professional home of the clinical elite, had long held. As a leader in the *SMJ* warned in 1962:

> Specialization ... has come to stay and the history of medicine suggests that the trend will continue. This being so it would be wise for us to recognize and to try to come to terms with it rather than sigh for 'the years that are past' [but] generalists wonder what becomes of the patient himself in these *sancta sanctorum* [specialist units]. Is anyone trained to recognize the pneumonias, myocardial infarctions, renal failures or any of the ailments that inpatients are prone to? Or is the patient simply regarded as a 'prostate' or a 'cerebral tumour' or 'neck of femur' or a 'bronchial carcinoma'. Who is the real doctor in a special unit? [15]

Addressing such debates involved a complete recasting of the very concept of medicine – and the final acceptance of the permeation of medical practice with scientific styles of thought. It did not mean that the specialists would not break away anyway. As Rosemary Stevens wrote in the middle of this crisis (in 1966) articulating these problems:

> Which was immediately more important to the young graduate: the general view or the swift acquisition of technique? ... Was a general grasp of optics more important to the ophthalmologist than a general grasp of other medical specialties? Was it more beneficial to generalize from a specialty or to specialize after a postgraduate period of general experience and, in the specialized world of the hospital, how was general experience ideally to be achieved? [16]

The situation cried out for a thorough reappraisal of what training at which point in the career ladder was required for which doctors; how this

[14] Gayer Morgan, *BMJ*, 2 (1963), p. 743. Cited from Rosemary Stevens, *Medical Practice*, p. 348.

[15] 'The Specialists', leader in the *SMJ*, 7 (1962), p. 49.

[16] Stevens, *Medical Practice*, p. 348.

was to be squared with the demands of NHS work; and what the relationship of the qualifications of the ancient medical corporations was to the demands of modern, specialised medical practice. A new conception of medical education, no less, was needed to train modern hospital doctors effectively. After graduation and the pre-registration year, the progress to consultant status had to be formalised, organised and standardised on a national and local basis. To do this there had to evolve at least a working definition of what a consultant actually was, and how each advance towards this was to be assessed qualitatively. As the GMC themselves importantly recognised in 1966, referring to the increasing pressure they were under to take some formal role in postgraduate education akin to that they already performed in the undergraduate sphere:

> The time is approaching when it will be necessary to identify schedules of training for each of the vocational branches of medicine (including general practice) and to arrange by an extension of the Medical Act for some form of official recognition to become available (such as the inclusion of names in a vocational register) for individual doctors when they have satisfactorily completed the training schedules for their selected careers in medicine.[17]

In Glasgow, the new progressive chiefs (many of whom had been appointed by Hetherington) were continuing to facilitate the development of specialisation in the hospitals. Further work is needed to clarify the details of this story and only a few brief examples can be given here to indicate the general trend. Illingworth, for instance, encouraged the development of neurosurgery and neurology at the GWI after Sloan Robertson had been appointed surgeon to the neurosurgical clinic in 1941. Illingworth also fostered the development of orthopaedics under Roland Barnes as University Lecturer from 1943. Barnes had been with Harry Platt in Manchester. Both these specialties were developed initially at the wartime Emergency Medical Service hospital at Killearn, by these GWI staff with the full support of Illingworth. In 1947 Illingworth obtained orthopaedic wards at the GWI for Barnes, who, in 1959, became the first holder of the Chair of Orthopaedic Surgery. Sloan Robertson continued to work at Killearn and an Institute of Neurological Sciences was established there from 1966, when Illingworth could not manage to persuade the Western to make room.[18] Ian Donald, appointed the new Regius Professor

---

[17] GMC, *Draft Recommendations as to Basic Medical Education* (1966) cited from Scottish Post-Graduate Medical Association, *Report of a Working Party on Postgraduate Medical Education within the National Health Service in Scotland*, p. 24.

[18] The Institute moved to new purpose-built premises at the Southern General Hospital in 1970. See Alistair Paterson, 'James Sloan Robertson, 1905–1975: An Appreciation', *College Bulletin*, 25 (1996), pp. 12–14.

of Obstetrics and Gynaecology in 1954, was also given beds in the Western, and got a new hospital (the Queen Mother's) for obstetrics in 1964. Another of Hetherington's appointments, Donald pioneered the use of ultrasound and laparoscopy. MacQueen and Kerr made the telling observation that he was (another) one, 'to whom must be given the credit for opening the eyes of the profession in Glasgow to the application of scientific methods in his special departments'.[19] In the early 1960s Illingworth was pushing for a new organisation of hospital work. He wanted to change his Department into a Division of Surgery to increase the influence of the academic unit and to break down isolating barriers between units. At the same time the University's new curriculum, and especially the new (fourth) integrated year aimed to break down barriers between medicine and surgery before they had formed too rigidly, and to expose undergraduates early to the clinical applications of scientific knowledge and techniques, was adopted. Both initiatives were responses to increasing specialisation.[20] At the Royal, too, specialisation was developing. In medicine this was first encouraged by Davis, and then by Edward McGirr as successive Muirhead Professors of Medicine, and also by J. H. Wright in his cardiological unit.

The younger Glasgow hospital practitioners (like Gavin Shaw, Robert Fife and McGirr himself) were part of the transitional generation between the older, largely part-time consultants and the coming full-time, more specialised, more scientific and research orientated generation. From these younger men, who had had to develop a 'special interest' to supplement their more generalist orientation,[21] came the demand for systematic postgraduate

[19] See MacQueen and Kerr, *The Western Infirmary*, pp. 156–61. Malcolm Nicolson of the Wellcome Unit is currently completing a book about Donald's development of ultra-sonography in Glasgow.

[20] See Sir Charles Illingworth, 'The Future of the General Surgeon', *SMJ*, 7 (January 1962), pp. 1–3; J. H. F. Brotherston, 'The New Curricula', ibid., (July 1962), pp. 291–95; and C. M. Fleming, 'The New Medical Curriculum in the University of Glasgow', ibid., (August 1962), 333–40.

[21] As ever, it is probable that Scotland had a distinctive take on specialisation. Because teaching hospitals were included under the RHBs (in contrast to England where they were independently governed), they had to undertake a larger share of routine NHS service work. This meant units continued to admit many general medical and surgical cases and that, consequently, specialisation may have occurred at a slightly slower pace than south of the Border. In addition, the Scottish tradition in medical education (which had long included a university element) may have meant that a more generalist outlook in which science and the clinical art were equally valued persisted at least until the generation considered here. See Robert Fife's contribution to the debate on 'The Place of the Regional Hospital in Undergraduate Teaching and the Training of the Consultant', in 'Royal College of Physicians: Meeting of Members', 24 November 1960, pp. 9–15, Shaw Papers; interview by Andrew Hull with Professor Edward McGirr, 28 October 1998, RCPSG, passim.

education, rotational schemes and standardised examinations. They were determined that future generations of Glasgow hospital practitioners would not have to waste time taking multiple higher qualifications and assembling positions and training themselves, aided only by their own chiefs, to form a coherent preparation for a particular career path. As much as their progressive seniors, they wished to see postgraduate education standardised and formalised to reflect the specialisation of hospital medicine. Many of these younger men (like Shaw, McGirr and Thomas Thomson) became heavily involved in postgraduate organisation, partly for the above reason, and partly because the challenge of teaching was useful to their own intellectual and professional development.[22] This generation had often gained wider clinical experience during their war service, and realised that the coming generation whom they were teaching needed a formalised version of this general experience. As Professor McGirr recently commented:

> We took a greater interest ... in the actual training of the junior staff in a clinical setting. In Davis's time many of us had had extensive experience, if not training, in the War, so he was fortunate in that sense. Later we had actually to start seeing that those with experience taught those lacking experience. It was the beginning, I suppose, of organised in-service postgraduate training.[23]

The next generation would take a single higher qualification after general professional training, then go on to organised specialist training. They would work in teams both in their service work and in their research. It was thus clear to both chiefs and rising juniors that the whole field of the medical education of the graduate must be reformed: advancing medical knowledge had to be synchronised with the training and service role of the NHS; and the higher examinations and the role of the medical corporations must be tailored to the new conditions.

These were the kind of problems that occupied the medical profession during this decade. The Todd Report of 1968 expressed the general consensus that had then evolved within the profession. The report established the mainstream for the future of postgraduate (and undergraduate) medical education, into which the Glasgow College then moved in collaboration with its sister Colleges.

At a meeting in Edinburgh of the standing joint committee of the three Scottish royal medical corporations on 15 January 1960 the subject of specialist diplomas was once again debated. The meeting expressed the opinion that

[22] See interviews by Andrew Hull with Gavin Shaw, Robert Fife, Sir Thomas Thomson, Professor Edward McGirr, Professor Arthur Kennedy, Robert Hume and Sir Andrew Watt Kay (and others) 1997–98, all held at the RCPSG.

[23] Interview by Andrew Hull with Professor Edward McGirr, RCPSG, 28 October 1998.

this was part of the function of the corporations, and that, 'the setting up of Diplomas might fuse together some of the splinter groups which would then be bound to the Colleges by their Diplomas'.[24] The debate seemed not to have moved on from the preceding year. Diplomas were still being considered as one possible way of dealing with the specialist problem, even though the number of different diplomas available in the UK was already very large and they were not considered by Appointments Committees, or by the specialists themselves, as signifying a very high standard of attainment. They certainly did not enjoy anything like the status of the Membership examinations. However, Glasgow's representative at the meeting, Joseph Wright, introduced a new note into these proceedings when he observed that:

> he felt that *bigger issues* were involved and that one should look to the future when it is possible that Statutory Postgraduate Higher Qualifications would become necessary and even required by Government instruction. *There will un-doubtedly be need to regularise qualifications in the future* ...[25]

The meeting agreed and resolved to seek contacts with the English Colleges to discuss these matters. Such cooperation was to be an increasing feature of the way the UK medical corporations worked from the 1960s onwards, leading to dramatic changes both in qualifications and in the role of the corporations.

Back in Glasgow, in a year when the Faculty was once more graced by a royal visit,[26] the more immediately pressing concern was the *modus operandi* of the GPGMEC, and especially the relative role and status of the Faculty and University. In Council on 16 February Shaw noted that communication between the Postgraduate Committee and the Faculty Council wasn't always satisfactory. There had been a specific recent problem over the organisation and financing of one of the ongoing series of conferences held in the Faculty, but things were now so bad that Shaw was going to have to ask the University's Director of Postgraduate Education, Charles Fleming, to ensure that copies of the committee's minutes were regularly sent to the Council. Three memoranda were produced (by Shaw, Wright – the Honorary Treasurer, and Goodall, as Vice-Convenor of the GPGMEC) for the next Council

---

[24] *Council Minutes*, 19 January 1960 (attachment), 'Minutes of Meeting of Standing Joint Committee of the three Royal Scottish Medical Corporations Held on 15 January 1960', RCPSG, 1/1/4/20 (Amalgamated Minutes, 1960–63).

[25] Ibid. (My emphasis.)

[26] Princess Alexandra of Kent was presented with the Honorary Fellowship of the College on 1 June 1960 in Faculty Hall. That evening she returned and dined with Fellows. In 1961 the College unveiled a splendid portrait depicting the Princess in her Honorary Fellow's robes by Sir Stanley Cursiter. This was unfortunately irreparably damaged in an unprecedented act of vandalism in July 1979. The current portrait in the Alexandra Room is on loan from P & O.

meeting, which was to be a special discussion of postgraduate organisation. The GPGMEC was a sub-committee of the University Medical Faculty. All Scottish universities had been requested to form such committees in 1944 at the request of the Department of Health for Scotland, and to coopt such local experts as they thought necessary to the provision of postgraduate educational programmes, in the first instance for demobbed doctors.[27] Glasgow University invited the Faculty to take part and it got four members to the University's eight, although as Goodall now commented, 'This distinction has never been of great importance in a Committee with a common purpose for most of the University representatives are Fellows of the Faculty and most of the Faculty representatives have some connection with the University'.[28] The Convenor was the University's Director of Postgraduate Education, and the Vice-Convenor represented the Faculty. During the 1950s, as we saw in the last chapter, the Faculty representatives asked for a substantial degree of freedom to establish experimental postgraduate courses. The Faculty had monies which could only be spent in this area and (this fact had helped it to realise that) it was clear that its future lay in postgraduate education and examination. The University readily agreed, busy as it was with undergraduate education and some routine postgraduate lecture courses, and on this basis the lectures, conferences, Ward Round and Recent Advances courses were begun. Shaw tactfully expressed the desire for some shift in the balance of power between the University and the Faculty in his memorandum, noting that:

> as a responsible Officer of the Faculty I am convinced that the future of the corporation, as of all the Royal corporations, lies in the development of Postgraduate medical education ... I feel the Glasgow Postgraduate 'School', if one could call it that, is beginning to take root and show real signs of life. In order to ensure that it flourishes it should be carefully nurtured. We are reaching a stage in which development, perhaps reorganization and reappraisal, of the supervisory and financial structure existing between ourselves and the University requires some consideration.[29]

The Faculty was clearly feeling that it did not have a status on the Committee consummate with the sheer amount of work it did in organising postgraduate activities, and that this had been underlined by recent events. However, there was also the problem of money. Figures presented to Council

[27] See Chapter 3 above.
[28] A. L. Goodall, 'Memorandum on Postgraduate Medical Education Committee', 9 March 1960, in RCPSG, 1/1/4/20.
[29] Gavin Shaw, Memorandum on Present State of Postgraduate Medical Education and its Organisation in Glasgow', 9 March 1960, Shaw Papers, RCPSG, 44.

by R. B. Wright, the Honorary Treasurer, showed that Faculty expenditure on postgraduate education had rapidly increased in the last years of the 1950s, until it was way beyond the annual £1000 contributed by the University. The shortfall had had to be made up from the trust funds to such an extent that the Lock Fund no longer had a surplus which could be channelled in this direction, since all available funds were now taken up by paying for the Lock Fellowship in Venereology at the University. Moreover, in 1960, the Faculty was running at a deficit of £867 9s. 9d.[30]

Table 5. *RFPSG Postgraduate Income and Expenditure, 1956–59*[31]

	Income	Total Estimated Expenses	Expenditure Charged to Trust Funds			
	£	£	£	£	£	£
			Fleming	Walker	Lock	Total
1956	1000	650	180	25	—	205
1957	1000	1375	200	200	375	775
1958	1000	1555	225	230	300	755
1959	1248	2030	250	150	750	1150

Later in his memorandum, Shaw was more explicit about the current problems that the Faculty had with the GPGMEC:

(a) It is a University Committee on which the Faculty is represented.

(b) Its system of communication with the Faculty is not defined.

(c) It is psychologically unfortunate to head advertisements with any organisation known as a 'Committee'.

(d) The role of the Committee is very far from clear. Is it legislative, advisory or executive body? What jurisdiction does it have, e.g. over the Faculty or individual employees in hospitals?

(e) Its financial arrangements are completely unknown and the general financial structure of the whole effort requires to be reviewed, regularised and controlled.

(f) Individuals on the Committee are able to take executive action on its decisions without further reference, at any rate to the Faculty.[32]

Shaw suggested a new postgraduate school or board with equal representation from University and Faculty, but with the University Postgraduate Director as permanent Chairman and three year periods of office for members. The new body should have the power to frame the policy for local

---

[30] *Council Minutes*, 7 November 1960.

[31] R. B. Wright, 'Memorandum on Financial Aspects of Postgraduate Activities', 9 March 1960, RCPSG, 1/1/4/20.

[32] Shaw, 'Present State of Postgraduate Medical Education'.

postgraduate education, it should plan finances, coordinate courses, publish a syllabus and place advertisements.[33] Goodall also stressed that it would be crucial that any new administration should attempt to attract postgraduate students from outwith Glasgow as well as serving the local needs, since, 'as Professor Graham consistently pointed out, "we cannot build a Postgraduate School by taking in each other's washing"'.[34] At the Council meeting on 15 March it was decided to propose to Charles Fleming (who had succeeded Wishart in 1959 as Dean of the Medical Faculty and Director of Postgraduate Medical Education, and who had recently been made Professor of Administrative Medicine to give him increased authority in dealing with other professorial members of the Medical Faculty),[35] that a joint meeting be swiftly arranged to discuss changes in postgraduate administration.[36] The Faculty was confident that Fleming would lend a sympathetic ear to its grievances since the special committee of the University Medical Faculty that had supported his appointment had included key Faculty men who were also Medical Faculty Professors (Cappell, L. J. Davis and Alstead). This committee had argued to the Medical Faculty in late 1958 that:

> We think it is important that there should be the maximum degree of cooperation in the field of Postgraduate education between the University and the Royal Faculty; and we hope that, while the Dean remains Chairman of the Postgraduate Committee, it may be found practicable to delegate as much as possible of the day-to-day Postgraduate work to officers of the Royal Faculty, who have in fact been undertaking a considerable proportion of it in recent years; it may even be possible to formalise this link between the University and the Royal Faculty.[37]

[33] Ibid.

[34] A. L. Goodall, 'Memorandum on Postgraduate Medical Education Committee'.

[35] Charles Mann Fleming (1904–1985) After taking an arts degree at Glasgow in 1924, he also graduated MB, ChB in 1929 and MD in 1933. He became a Fellow of Faculty in 1959, interestingly after taking up his position as Medical Dean and Director of Postgraduate Medical Education. He had previously (from 1946–59) been Principal Medical Officer to the Department of Health for Scotland, so he brought to his new job considerable experience in health administration. In 1970, when the title of his post changed as preparations were made for the 1974 NHS reorganisation, he became the first Postgraduate *Dean*. As Postgraduate Director and Dean he was *ex officio* Convenor of the Glasgow Postgraduate Medical Board, and, after 1970, of its successor, the Western Regional Committee for Postgraduate Education. He was awarded the CBE in 1964. See his Obituary in the *BMJ*, 291 (1985), p. 148, 13 July 1985, and *Who Was Who, 1981–90* (London, 1991). Fleming was one of Hetherington's last appointments. As usual he had taken a very close interest and had from the start assured Fleming that he would get professorial status so that he could continue, unhindered by academic prejudices, where Wishart had left off. See Professor Thomas Ferguson to Fleming, 8 August and 1 September 1958, GUABRC, DC119 8/3.

[36] *Council Minutes*, 15 March 1960.

[37] Glasgow University, *Medical Faculty Minutes*, 4 November 1958, GUABRC, DC119 8/3.

Shaw wrote to Fleming to initiate a meeting, outlining the Faculty's general plans which he hoped Fleming would put to the Medical Faculty. He stressed that the Faculty, 'had steadily been increasing its role in this field by supplying ideas, financial resources and administrative facilities', as well as the new lecture room and planned new study rooms. Shaw suggested a new joint organising board composed of, 'senior members of both Faculties', and tactfully stressed the 'spirit in which [the proposals] are offered, namely one of anticipation of greater future cooperative efforts in a growing enterprise'.[38] The two Faculties met on 21 April and again in early May 1960.[39] Meanwhile Alstead had let fly a very frank letter to Shaw, inspired he said by the opinions of the Visitor (J. H. Wright) on the matter. He had headed the letter: 'This is a personal letter – though I am willing to have it read to Council and Office Bearers at RFPS. As I indicate below, I am, of course, expressing merely a personal viewpoint':

> The time is ripe for the *University Court* and the RFPS *Council* (or their senior Office Bearers) to get together soon in order to discuss a constructive policy for Postgraduate medical education in the West of Scotland. I am confident that if the RFPS makes it clear that they mean business *in a big way* the University will be obliged to take this matter seriously. As I see it, what is needed is a vigorous attack on what I must call (without offence) a Laodicean attitude in academic circles; and simultaneously to destroy the myth that the RFPS is the poor relation living in a room and kitchen in St Vincent Street and needing an annual tip of £1000. The RFPS has earned the right to dominate this kind of graduate work- though it would be calamitous to insist on exercising that right. Once the correct atmosphere has been achieved I think that such matters as designation ('Board' instead of Committee) and equality of representation will fall naturally into place.[40]

As Professor of Materia Medica, Alstead, as Shaw recently remarked,[41] was more privy than most to 'Laodicean' mutterings at the University about the desirability or capabilities of the Faculty taking a 'dominant' role in post-graduate education. The background to Fleming's 1959 appointment would seem to indicate that the University bureaucracy and at least some important professorial members of the Medical Faculty were still as keen as under Hetherington and Wishart to allow the Faculty a fair degree of autonomy in postgraduate matters. However, that may be to miss Alstead's point. He may simply have been expressing the frustration felt in Faculty at being technically subservient to a sub-committee of the University Medical Faculty.

[38] Shaw to Fleming, n.d. but early April 1960, Shaw Papers.
[39] *Council Minutes* 21 April, 17 May 1960.
[40] Alstead to Shaw, 22 April 1960, Shaw Papers. (Original emphasis).
[41] Gavin Shaw, interview with the author, 1997.

None of the other Royal medical corporations was in this invidious position. The Faculty had now outgrown this arrangement (though it still needed the University's money). The detailed agenda/memorandum for the May 1960 joint meeting formally proposed a Postgraduate Medical Education Board, a new split of work, and contained a request from the Faculty for an increase in the annual grant from the University from UGC funds to between £2500 and 3000.[42] The University approved the new arrangements without further ado, although the grant was only increased to £1500, and the respective representation remained at Faculty four; University eight. However, once again this was really not an issue. As a glance at the initial membership of the new Glasgow Postgraduate Medical Board (GPGMB) immediately reveals, the 'Faculty' representatives were as likely to be aware of University issues as the 'University' representatives were to be aware of 'Faculty' perspectives.[43]

One other initiative undertaken during 1960 that was later to bear important fruit was the first foray into the question of the change of name. When the President, Arthur Jacobs, had mentioned the Faculty's intention to seek a change of name to the 'Royal *College* of Physicians and Surgeons of Glasgow' to the July joint meeting of the Scottish medical corporations, the (rather insulting) response had been to enquire of him whether this was meant simply as a prelude to the union of the Scottish Colleges. At the special Council meeting called to discuss the matter on 6 September, Jacobs spelled out why such a name change was imperative:

> the title of the Faculty left the Royal Faculty at a disadvantage as all the sister Corporations were known as Colleges. Of recent years the term Faculty was being used increasingly for bodies of specialists within the Colleges but of definitely lesser status. Some such Faculties might even become Royal Faculties.[44]

The meeting agreed to seek the opinion of the Faculty's solicitor (T. L. Grahame Reid) in this complicated matter, which might require an expensive private Bill to be raised, and would certainly mean careful negotiation with the sister corporations over the titles of future higher qualifications.

By 1961 the GPGMB was now firmly established and functioning as the

---

[42] 'Memorandum for Joint Meeting between the Royal Faculty and University on Proposed Postgraduate Board, May 1960', Shaw Papers.

[43] Faculty Representatives: J. H. Wright, Charles Illingworth, Professor Roland Barnes, Professor D. F. Cappell. University Representatives: Professor Ian Donald, Dr E. M. McGirr, Professor T. Symington, Professor William Arthur Mackey, Professor E. J. Wayne, Professor T. Ferguson Rodger, Dr J. H. Hutchison and Dr C. M. Fleming. List from 'Postgraduate Medical Education Committee', Shaw Papers.

[44] *Council Minutes*, 6 September 1960.

central coordinating body and information bureau for all postgraduate educational activities in the region. From September the board was producing
a monthly 'medical calendar' of all local meetings and courses.[45] The Faculty's
surgeons were expressing the desire that the hospitals and departments in
which candidates for the Fellowship received their training should be increasingly supervised and selected by the Surgical colleges acting together.
Though this was done by a Faculty sub-committee in terms of the regulations
for the Final Fellowship, joint action might better ensure the adequacy of
training posts.[46] In Dundee, at a meeting of Scottish Medical School Deans
and representatives from the royal medical corporations, it had been agreed
to form a Scottish Postgraduate Medical Association (SPMA), composed of
university and royal medical corporation representatives, to promote exchange of ideas and work towards regional standardisation of courses. The
new body first met at the RCPEd on 15 December 1962.[47]

By far the most important event of 1961, however, was the conference on
postgraduate medical education sponsored by the Nuffield Provincial Hospitals Trust (NPHT), and held at Christ Church, Oxford, on the 16–17
December under the chairmanship of George Pickering, the Regius Professor
of Medicine at Oxford. The NPHT and the Oxford University Medical
Faculty, which had been heavily involved in the preparation of the seminal
Goodenough Report in 1944,[48] which had set the tone for the future reforms
of undergraduate medical education noted above, were now also to be
influential in the modernisation of postgraduate education and training. The
NPHT had managed to draw together an impressive group of delegates
representing the Ministry of Health, the UGC, the Universities, the Royal
Colleges, the British Postgraduate Medical Federation, boards of governors
of teaching hospitals, Regional Hospital Boards and provincial consultants.
Charles Fleming attended for Glasgow. The conference first of all agreed on

[45] See Minutes of the Glasgow Postgraduate Medical Board, 20 September 1961, Shaw Papers.

[46] Minutes of a 'Meeting of the Surgical Councillors to Discuss Points Remitted from the
Meeting of the International Federation of Surgical Colleges', 3 February 1961, RCPSG, 1/1/4/20,
Amalgamated Minutes, 1960–63.

[47] 'Minute of Meeting of the SPMA Provisional Council', 20 October 1962, RCPSG, 1/1/4/21.
The Glasgow College got one Medical and one Surgical representative, as did the other Scottish
Colleges. Universities also got two representatives (one of whom was the Director of Post-
graduate Education), and there was also one observer from the Scottish Home and Health
Department. See, 'Minutes of Joint Standing Committee of the Royal Scottish Medical Cor-
porations', 28 April 1962, RCPSG, 1/1/4/20.

[48] See Nigel Oswald, 'A Social Health Service without Social Doctors', *Social History of
Medicine*, 4 (1991), pp. 295–315; idem, 'Training Doctors for the National Health Service: Social
Medicine, Medical Education and the GMC, 1936–48', in Dorothy Porter (ed.) *Social Medicine
and Medical Sociology in the Twentieth Century* (Amsterdam, 1997), pp. 59–80.

fundamentals: 'It is of urgent and immediate concern for both the present and the future of the health services that the pattern of, and facilities for, education in the hospitals of the regional boards should be radically improved'.[49] The Spens training-ladder must be returned to and made the basis for NHS staffing policy so that, 'All posts ranging from pre-registration appointments to those of Senior Registrars should be recognised as training posts'.[50] Pickering explained why this re-emphasis was acutely necessary at the present time in early 1962. In spite of recommendations to the contrary in both the Spens and Platt Reports:[51]

> This particular situation has arisen because the junior posts of House Officer, Senior House Officer, Registrar and Senior Registrar are the means by which much of the medical work of the service is done. Their number has not been exactly related either to the needs of education or to the probable vacancies at consultant or other levels.[52]

Pickering was talking about the English situation. In England the teaching hospitals were run independently of the RHBs as primarily undergraduate teaching institutions, while in Scotland they were included in the RHB administration and thus also did a larger amount of NHS service work. Pickering argued that, in England, this dual organisation had meant that the large regional district hospitals had been seen as service and not as teaching centres. With the above caveat, however, his point was equally valid for Scotland. The majority of young hospital doctors were employed in district hospitals, which, given the undergraduate responsibilities of the teaching hospitals, were not extensively used for *systematic* postgraduate education, though their units did provide a limited number of highly prestigious and sought after training posts. This meant that the natural home for the bulk of postgraduate education and in-post training – the district hospital – was not prepared for the task. The conference recommended that the district hospital or group should now be recognised as the basic regional postgraduate training unit. Here an 'educational atmosphere' should be promoted and time should be made available above service commitments for junior and senior staff to devote to education.[53]

[49] 'Postgraduate Medical Education: Conference Convened by the Nuffield Provincial Hospitals Trust', anonymous article in *Lancet* (1962), i, pp. 367–68.

[50] Ibid.

[51] *Report of the Joint Working Party on the Medical Staffing Structure in the Hospital Service* (London, 1961), the Platt Report. Chaired by Sir Robert Platt, President of the RCPL.

[52] Sir George Pickering, 'Postgraduate Medical Education: The Present Opportunity and the Immediate Need', *BMJ*, 17 February 1962, pp. 421–25.

[53] Unless indicated this and subsequent short quotations are from the *Lancet's* report of the Christ Church conference cited above, on which this section is largely based.

In Hetherington's Glasgow, as we briefly noted earlier, some of the district hospitals (notably the Southern General) had been prepared for this post-graduate role from the mid 1940s. During the war, the Southern General had been seeded with (university-affiliated) consultants from the major teaching hospitals with the cooperation of the University and the City Coun-cil (as had happened with Stobhill), juniors were also appointed to these men. Thus certain Glasgow 'peripherals' had been upgraded even before their take-over by the NHS RHB in 1948.[54] The Christ Church conference went much further, arguing for a thorough upgrading of the postgraduate status and facilities of these hospitals. Each district hospital or group should have a 'clinical tutor' nominated by the regional postgraduate committee. He would be responsible for teaching arrangements and career guidance for all those in training. He should also be responsible for contact with local general practitioners to allow them equal access with trainee consultants to the hospital in the clinical discussions, clinico-pathological conferences, li-braries and diagnostic departments that were to be established or upgraded. The regional postgraduate training unit should be fully equipped with a seminar room, library (and part-time librarian), clinical tutor's room, and laboratories adjacent to the wards in which trainees could perform tests on their patients. Academic style standards were to be adhered to in supervision, quantity and variety of clinical material, records, post-mortem service, X-ray and pathological services, laboratory facilities and services and clinico-pa-thological conferences.

While the units were the basic elements in the new regionalised structure of postgraduate education, there should also be, 'an association with the regional university'. The University Postgraduate Dean or Director, ap-pointed by the Medical Faculty, was seen as the local linch-pin, 'on whom the success or otherwise of this scheme will very much depend'. The Dean was to have ultimate responsibility for the interrelationship between the regional hospitals and the university; the arrangements for teaching in the region; career advice; advising and placement of overseas graduates; and the integration of GPs into the postgraduate arrangements. He should work in conjunction with a 'strong Regional Committee for Postgraduate Education' which would have representation from the regional university, the RHB and the relevant royal medical corporations. He would also be responsible for the general coordination of the scheme, for rotational appointments schemes, and the appointment of clinical tutors. The university was to pay the Dean and the clinical tutors, and the RHB to pay expenses incurred by their staff. While recognising that immediate action hinged on local initiatives, the

[54] Interview by Andrew Hull with Gavin Shaw, 29 October 1997.

conference urged the formation of a national body (with UGC, Universities, GMC, Departments of Health, Royal Colleges and Corporations and the College of GPs represented on it) to be responsible for, 'central formulation of principles and of policy'.

The Nuffield made £250,000 available as start-up finance to begin the regional implementation of this scheme for the organisation of postgraduate medical education. Regions were invited to apply competitively for a share of this money. The funding announcement stressed that: 'The trustees will be *specially* concerned to see that the appropriate university and regional hospital authorities are both wholeheartedly behind any approach to the Trust ...' [55]

The Nuffield was thus extending its continuing active encouragement of a national policy of both regionalisation and academicisation of medical education in this new scheme. It was attempting to push towards a national policy (either voluntary or legislative) on postgraduate education that saw it centred in the regional hospital structure that it had itself helped to make the basis of the NHS. Postgraduate education would revolve around the local university, which had been the main thrust of the arguments of the Nuffield influenced Goodenough Report for the reorganisation of undergraduate education which had had such an impact nearly twenty years previously. Now Glasgow was well ahead of the game. Much of the Nuffield's suggested postgraduate infrastructure was already up and running in the western region. Gavin Shaw has described the Christ Church conference as a 'benchmark'. During 1962 the GPGMB appointed twenty-one clinical tutors (later known as postgraduate advisers) to Glasgow's hospitals. [56] The funding for these positions was taken over by the RHB (with funds from the DHS) in 1964, a recognition both of the national importance of postgraduate medical

[55] 'Support from Nuffield Trust', *Lancet*, 17 February 1962, p. 368. My emphasis.

[56] The twenty-one clinical tutors originally appointed in early 1962 were: Glasgow Royal Infirmary and associated hospitals, Dr A. C. Kennedy and Goodall; Glasgow Western Hospitals, Mr D. H. Clark and Dr I. A. Short; Glasgow Northern Hospitals, Dr T. J. Thomson; Glasgow Victoria Hospitals, Dr T. Semple; Glasgow S. W. Hospitals, Dr G. B. Shaw; Glasgow and District Children's Hospitals, Dr Eric Coleman; Glasgow Royal Mental Hospital, Dr T. Freeman; Glasgow Women's Hospitals, Dr J. M. McBride; Northern Ayrshire Hospitals, Dr G. M. Currie; Southern Ayrshire Hospitals, Dr A. Slessor; Dumfries and Galloway Hospitals, Dr J. Laurie; Crichton Royal Mental Hospital, Dr A. C. Tait; Coatbridge, Airdrie and District Hospitals, Dr R. A. Tennent; Motherwell, Hamilton and District Hospitals, Dr H. R. F. Macdonald; Southern Lanarkshire Hospitals, Dr A. Mackenzie; Paisley and District Hospitals, Dr H. Conway; Greenock and District Hospitals, Dr J. Fleming; Stirling and Clackmannan Hospitals, Dr A. A. MacKelvie; Falkirk and District Hospitals, Dr R. G. Henderson. See 'The Development of Postgraduate Medical Education' (n.d. but probably May 1962), Shaw Papers, RCPSG, 44.

education, and of how the Nuffield-sponsored scheme was (once again) evolving into the national policy. In 1965, Shaw noted the importance and the impact of these appointments:

> Now, three years later, there has been the most staggering upsurge of activity throughout the country. Area postgraduate centres are appearing where no flicker of educational activity had ever burned before. Regional directors are being appointed and hospital advisers springing up. But most of all, the Government has acknowledged its responsibility, and, far from being a dirty word, postgraduate education has become a most important aspect of the NHS. These men [clinical tutors and postgraduate advisers] are given University status and a small honorarium and act as liaison officers with the centre and with the local College of General Practitioners. This has been a great success in the peripheral hospitals resulting in useful and well-attended meetings and clinico-pathological conferences. In the centre the Postgraduate Advisers have had the same job as well as caring for many overseas students on clinical attachments. This knitting-together of the whole under the control of the Postgraduate Board has worked well. The Royal College has provided library, reading rooms and photocopying to all postgraduate students in the West of Scotland.[57]

Under the previous *ad hoc* postgraduate arrangements, trainees consultants had had to map out their own career paths, going where there was a suitable position. For example, Arthur Kennedy went to Dumfries for a Registrar appointment. Here he obtained good experience, and he returned to the GRI to a lectureship in Leslie Davis's University Medical Department. Not all the peripheral hospitals were up to this standard:

> There were other places where to be sent made no difference at all. The senior staff weren't interested, were shy of teaching; they hadn't the organisation. But once they got a good Postgraduate Adviser that made all the difference, he could stimulate things, and they did this by inviting a lot of guest lecturers down.[58]

In short, postgraduate advisers, holding honorary university status and, initially, paid by the university for their work, helped to spread the academic ethos of interlinked teaching and research within a more scientific and specialised concept of medicine that Hetherington's appointees had brought to the Glasgow teaching hospitals, from these central teaching hospitals to the peripheral or district hospitals which were now to be increasingly used for the postgraduate education and training of consultants. However, postgraduate training was so unorganised in many areas that the Nuffield's

[57] Gavin Shaw, 'The West of Scotland and Postgraduate Education', paper presented at Portland, Oregon, conference on postgraduate education, May 1965, Shaw Papers, RCPSG, 44.

[58] Interview with Gavin Shaw by Andrew Hull, 29 October 1997.

favoured 'educational atmosphere' was not intended to be simply an academic one. Pickering stressed the crucial importance of familiarity with laboratory methods (and praised Scottish medical education for still inculcating this with the result that most hospital doctors educated there would do their own side-room tests and not constantly send to the laboratory), but he also stressed the importance of the traditional apprenticeship model of relations between juniors and their chief in the clinical setting. Pickering stressed that 'educational atmosphere' did not necessarily mean a research culture (as existed in the central teaching hospitals and was partly encouraged by the presence of academic units), but a 'culture of learning' of the clinical art:

> Most of us would agree that the most important formative factors in our lives have been the chiefs under whom we have served. The most important feature of a good [postgraduate] teaching unit will be the standard of excellence demanded by the staff from the Registrars and Housemen. If histories are first class, physical examinations accurate and complete, progress notes accurate and up-to-date; if the Registrars and Housemen themselves make simple examinations of body fluids, watch their patients being screened or operated upon, and regularly visit the post-mortem room – then it can be said that all the most important functions of the teacher have been performed.[59]

By stressing the continuing importance of the inculcation of traditional clinical skills, Pickering may have been playing to the clinical elites of the regions where he wished to upgrade postgraduate training, who were nervy about the extension of academic influence into this new field.

In Glasgow such problems had already largely been solved. Here the existing (post-1960) new postgraduate organisation, with its more equal and harmonious relations between University and Faculty, continued. However, it was clear that the Faculty's drive for more efficient and fair organisation had been strengthened by the imprimatur of Christ Church. During 1962 the GPGMB made their submission to the Nuffield Trust for a share of the £250,000. They pointed out at the start that postgraduate numbers had risen from a mere 150 in 1957 to 1240 in 1961.[60] The submission outlined the new division of work within the board. The board itself was responsible for general policy and finance and with the approval of preregistration House Officer posts. A postgraduate committee of younger consultants handled the planning and organisation of specialist courses, and the detailed organisation of

[59] Pickering, 'Postgraduate Medical Education', p. 424.

[60] 'Glasgow Postgraduate Medical Board: The Further Development of Postgraduate Medical Education in Glasgow and the Western Region. A Submission to the Nuffield Provincial Hospitals Trust, April 1962', GUABRC, DC 119 4/1.

specialist courses was delegated to *ad hoc* sub-committees to which other young specialist consultants were co-opted. Clinical tutors – most of whom were Glasgow University graduates and Fellows of Faculty – providing the channels of contact with all the postgraduate teaching units in the district hospitals, and handling the local execution of the board's programme. The joint University/Faculty membership of the board ensured, it was argued, 'continuity and integration with undergraduate education and [gave] to postgraduate education both impetus and authority'.[61] The linch-pin was the board's chairman – the Medical Dean and Director of Postgraduate Education. The board favoured the continuation of this combined appointment since it meant that coordination of under and postgraduate medical education was achieved through the one individual being responsible for both. As well as being chairman of the board, the Dean was also a member of the RHB and its Medical Education Committee, the Medical Staff Associations of both the GWI and GRI and the undergraduate and postgraduate committees of the West of Scotland Faculty of the RCGP. He also had monthly meetings with the Senior Administrative Officer of the RHB to discuss matters of common interest to the University and the regional board. Under the Director were now four Assistant Directors (consultants from the two main teaching hospitals) who assisted in coordinating the work of the clinical tutors and advising students. Courses for GPs were run by the Medical Staff Associations of the five teaching hospitals.

The Faculty's role was now explicitly formalised too: it was to become the 'postgraduate centre' around which the board's activities revolved.[62] It was ideally suited to take on officially the function it had been increasingly providing since the early 1950s with its central library service, weekly lectures for junior staff, preparatory courses for higher qualifications and sectional meetings along the lines of the Royal Society of Medicine open to hospital staff and GPs. As the submission argued:

> The Postgraduate Board believe that their activities would gain enormously in cohesion and value if based on a central premises. The RFPS are now concerned almost exclusively with postgraduate education and are prepared to offer the use of their very conveniently situated rooms and library for such development. The present time is most opportune. A central Library service could readily be developed. The Royal Faculty Library contains some 150,000 volumes and subscribes to the leading medical journals. An extension, costing over £20,000 and providing additional reading and other rooms, will be completed in the next few months ... The Royal Faculty are willing to allow all junior hospital medical staff in the

---

[61] Ibid.
[62] Ibid.

region to have access to the Library. Few measures would do more than a good central library service to promote the desired educational atmosphere for the Postgraduate Board's activities.[63]

From 1963, the Glasgow College obtained full responsibility for the running of conferences under the board, thus solving the initial difficulty which had been the proximate cause of its agitation for a more formalised role in postgraduate education. At that point the new Postgraduate Advisory Committee of the College also became a Conference Committee, initially with Shaw as its Convenor.[64]

The board sought nearly £50,000 from the Trust, but got £25,000, which funded the clinical tutors (until they were funded from the DHS in 1964). A large amount of money was sought, but not obtained, for an experiment in the employment of television in the education of doctors. The original idea was to transmit educational programmes from the teaching hospitals to the district generals, as part of an ongoing effort to bind the two together. When this funding was not forthcoming, the board linked up with Scottish Television. As in so many other aspects of the modernisation of the Glasgow Medical School, Sir Charles Illingworth was the prime mover. He had been inspired by the example of the BBC's programme 'Your Life in Their Hands', which was aimed at the public, to develop the use of television as an educational medium for doctors. As Gavin Shaw recently recalled:

> This was Charlie Illingworth's idea, as so many things were. He had, I think, been to dinner with Roy Thomson of Scottish Television and had got talking, and had twisted his arm to the extent that it might be a good idea if we could put a television programme on STV – it would give STV a better image. Thomson said, 'Well, I think it's a good idea, but I think we've got to try it out'.[65]

As part of a conference at the University in late 1962, STV, the Pye Group and Illingworth's University Department of Surgery collaborated to broadcast an experimental open-circuit medical teaching programme on haematemesis. The programme was produced in the GWI and transmitted on an open channel for viewing in the district hospitals. After this initial success, STV and the GPGMB began collaboration on a more ambitious series of twenty-six monthly programmes under the title 'Postgraduate Medicine' which were transmitted from March 1963, each programme being once repeated. The first programme was on renal disease. Our illustrations, showing all of the key individuals, are taken from the thirteenth programme, 'Peptic Ulcer: A Surgeon Cross-Questioned', in which, fittingly, Illingworth himself took the

[63] Ibid.

[64] *Council Minutes*, 16 April 1963, RCPS, 1/1/3/14.

[65] Interview by Andrew Hull with Gavin Shaw, 29 October 1997.

leading role.[66] The programmes, which were about twenty-five minutes long, went out late at night, after the regular service had closed down, but the board also received a 16 mm. sound copy for distribution, mainly to general practitioners.[67] A special committee of the GPGMB was formed consisting of two physicians, Gavin Shaw (Convenor) and Thomas Thomson; a pathologist, Bernard Lennox; and a surgeon, A. D. Roy. The talented medical artist Gabriel Donald was later added to help with problems of medical illustration. This series was the first in Europe, although there had been an experiment with televised postgraduate education for doctors in Utah in 1953. The success of the Glasgow scheme prompted the BBC to enter the field, and they launched their own programme called 'Medicine Today' in 1964 in conjunction with the Association for the Study of Medical Education.[68] Between 1963 and 1969 a total of seventy-three programmes were produced by the independent television team, although the later ones were mainly done with the Medical Schools of Newcastle, Edinburgh, Dundee and Aberdeen, and some were produced by Tyne Tees Television.[69]

As well as seeing the successful culmination of the Faculty's campaign to reform the organisation of postgraduate education in Glasgow, and to formalise a key role for itself in that enterprise, 1962 also saw the continuation of the debate on the nature, content and purpose of the Faculty's higher examinations, and the successful accomplishment of the change of name.

The rethink on postgraduate training and education also meant that the Faculty had to reexamine the role of its higher qualifications in medicine and surgery. If postgraduate training was standardised on a national basis all the royal medical corporations would have to reevaluate the purpose and content of the Fellowships and Memberships. What role would they play in denoting the trained specialist, and at what point on the career ladder would they be taken? In January 1962 the Faculty set up two committees to review the Fellowships in Medicine and in Surgery. The first question addressed was how to relate the conditions of entry more closely to the training ladder, and to the existence of more specialised units. The committee on the Final Surgical Examination recommended a more detailed breakdown of the mandatory three years of clinical experience in approved posts necessary before sitting. After the preregistration year, two years should be spent in

[66] Some of the earlier programmes still survive in the archives of the Scottish Media Group (formerly STV) in Glasgow. I am grateful to the librarian for this information and for permission to reproduce the accompanying images.

[67] Interview with Shaw.

[68] On this see A. D. Roy, 'Scottish Television Experiment', *BMJ* (1964), 2, pp. 1321–23; Gavin Shaw and Arlene Smith, 'Does Medical Television Work?', *Health Bulletin*, 28 (January 1970), pp. 1–6.

[69] 'Postgraduate Medicine: Subject Titles, March, 1963 to June 1969', Shaw Papers.

General Surgery, they suggested, one of which could be spent in research or teaching in an allied science. Also the third year could be spent in a range of specialties (orthopaedics, urology, plastic surgery, neurology, casualty surgery, paediatric surgery, otolaryngology, ophthalmology, gynaecology or a combination of these). Candidates intending to specialise in the examination in one of the three available options (Otolaryngology, Ophthalmology and Gynaecology) would need to complete the final year in an approved post in that specialty. The committee also recommended that the form of the Final Examination should change to reintroduce a general element: thus it was recommended that all candidates should once again take written and oral examinations in the Principles of Surgery, before opting to take written, oral and clinical in the practice of either Surgery, Otolaryngology, Ophthalmology or Gynaecology as a specialist subject. Interestingly, this meant that general surgery was also now being considered a specialty, which was later to become the generally accepted solution to the problem specialisation posed for general medicine and surgery: a case of all specialists now.

The committee which had been examining the Medical Fellowship recommended that, 'the Council should eventually adopt the Edinburgh pattern as the sole form of examination'.[70] This clearly shows how focused Faculty minds were on the question of specialisation. The Edinburgh Physicians had had a uniquely specialist Membership since the introduction of an examined Membership in 1881. One general set of examinations was taken, then one in an ever-expanding list of designated special subjects.[71] In contrast to the rather static Glasgow numbers in the early 1960s, the Edinburgh examination was very popular, particularly with overseas candidates who often took it to indicate full specialist status. In 1960 the Edinburgh exam had attracted 1252 candidates, in 1961, 1323 and in 1962, 1373.[72] The Edinburgh Final Examination differed in two respects to its current Glasgow equivalent. First, the third year of clinical experience had to be in the specialty. Secondly, whereas Glasgow had a generalist exam with written, oral and clinical tests in the Principles and Practice of Medicine, Edinburgh's Final offered two papers in General Medicine (with question choice, plus a clinical and an oral), and one set of examinations in the chosen specialty. Moreover, by 1962, Edinburgh offered a mind-bogglingly broad choice of specialties.[73]

The recommendations of both committees were not acted on just yet, but

[70] 'Report of a Committee Appointed to Consider the Examination for the Fellowship *qua* Physician', chaired by Alstead, May 1962, RCPSG, 1/1/4/21.

[71] Manderson et al., 'Evolution of an Examination'.

[72] Craig, *History of the RCPEd*, p. 750.

[73] Committee on Fellowship *qua* Physician, May 1962. Of the Edinburgh special subjects, the committee considered the following suitable for inclusion in a revised Glasgow examination:

demonstrate that the Glasgow Faculty was turning these questions over very thoroughly as early as 1962. Well it might, for so were the other corporations. A meeting of the Presidents of the Colleges of Physicians of London, Glasgow, Edinburgh, Ireland, Canada, Australasia and South Africa in October made the first moves towards reciprocity of final examinations.[74] The conference's 'Final Thoughts and Tentative Conclusions' indicated the mood and direction: all participants were agreed that four years' hospital training were ideally needed after graduation before sitting higher qualifications, and that all bodies should draw up lists of approved hospital posts to enable tighter control of the quality of training. The conference concluded that, 'The danger to traditional Colleges of "splinter-Specialties" is a real one, and there is therefore a strong case for recognition of specialties within the Colleges'.[75] However, extending the required training times meant that the point in the career at which a candidate took the examination was gradually creeping later and later. The question of whether the higher qualifications were going to be early general tests of potential to move on to specialist training or late tests of specialist status was soon to become increasingly pressing.

On 6 December 1962 the provisional order changing the Faculty's name to Royal College was promulgated. J. H. Wright had handled much of the detail on this project. Wright had vowed when he accepted the Presidency in 1960 that he would accomplish two specific things: to change the name to College to give it equal standing with its sister corporations, and to provide a much improved library for the College to make it a more viable postgraduate centre. Gavin Shaw recently shed some light on the wranglings over the first objective:

Joe [J. H. Wright] had the idea and it was left to me and one or two others to

Alimentary Diseases, Cardiology, Child Life and Health, Dermatology, Blood Diseases, Endocrinology, Geriatrics, Infectious Diseases, Neurology, Psychiatry, Respiratory Diseases and Tropical Medicine. The following were considered unacceptable: Anaesthetics, Anatomy, Forensic Medicine, Medical Ophthalmology, Obstetrics and Gynaecology, Experimental Pharmacology, Public Health and Social Medicine, Renal Physiology and Diseases, and Tuberculosis. The committee considered that there was a third group of Edinburgh subjects which might or might not be suitable, 'depending on matters of policy'. These were Bacteriology, Biochemistry, Pathology, Radiodiagnosis and Radiotherapy, Rheumatology and Physical Medicine and Venereal Diseases. The way that this list was divided by Glasgow reveals a lot about the Faculty's contemporary thoughts on what was a speciality, and what should rightly be considered part of general medicine. As ever, it also reveals a pragmatic willingness to go wherever the other Colleges might wish to go on this matter.

[74] Glasgow was still a Faculty, of course, and both it and the South African College comprised Physicians *and* Surgeons.

[75] 'Proceedings of Conference of Presidents of Physicians of the Commonwealth and Associated Republics at RCPEd, 4–5 October 1962', p. 7, RCPSG, 1/1/4/21.

find out how you did it. We went to see the College lawyer and he said that under Scots Law it was not necessary to get an Act of Parliament ... we got it through the Faculty, and we then went to the General Medical Council, and, boy, did we run into trouble. First of all they said we'd taken the wrong route in Parliament, there was no way that we could get this through without a full Act of Parliament. Secondly they said, this question of the names is a nonsense, we can't have this. We were very downhearted at that stage, we almost failed. So Joe and his friend Lord Fraser went down to London, and they invited a whole lot of senior GMC people out to a lunch party. They got a little further then, and Joe said I think I see the light at the end of it. The obstacle was the Registrar, who was an English lawyer, and very careful. So then, very fortunately, the then President of the GMC was the Professor of Materia Medica at Aberdeen, Sir David Campbell, a Glasgow graduate and FRFPS, and one of our Fellows. He was a friend of Joe's, and Joe got him and said, look this has got to be put right, our clerk is absolutely certain that we can do it by this route, which is a quick route, provided we've got everyone agreed. One wintry day we were all invited to his house in Aberdeen, so we all went up on the train, plus the lawyer. So Joe started off, and the President made the case for them. Our lawyer was a silent man ... so he sat and not a word was said while this Registrar put his case, and destroyed our case. Then the President said, well now what does your man have to say, and he said, 'You may be a very good English lawyer, but you don't know Scot's Law!', and that silenced him completely.

There were also administrative changes to be introduced into the provisional order, and an oversight to be rectified:

There then followed a long period when I started work to change the constitution, Archie [Goodall] and I wanted to put in a lot of changes, not just the name, but how we would elect Fellows and so on. I was going down to meet the Parliamentary draughtsman. And then, just as we were at the stage of thinking all was well, the Dental Convenor phoned me up and said, 'I know the Chairman of the Dental Council, and you haven't asked him'. So I phoned Joe and said, 'We forgot the GDC', and he said, 'So we have', in fact nobody had remembered it, and the Dental representative on the Council hadn't remembered it either. So the Dental representative said, 'Leave it to me and I'll sort it out', and he did in about ten days or so. So eventually it went down and the Parliamentary people did whatever it is that they do, and the next thing we got a letter saying that the Queen was about to sign it, and that was that. It was formally put before Parliament. There were no objectors. It took a lot of time, but Joe was the prime mover.[76]

After the GMC had agreed, the other Colleges also gave their approval quickly and 'graciously'.[77] Under the order, all existing and future Fellows

[76] Gavin Shaw, interview with Andrew Hull, at RCPSG, 29 October 1997.

[77] Joseph Wright, transcript of interview with Peter MacKenzie, RCPSG, 8 February 1985, RCPSG, 18/20.

qua Physician became Members, with the title MRCP (Glasg.). A select group were automatically elected to the now senior grade of Fellows *qua* Physician. This title was now conferred on the recommendation of the Fellowship Committee appointed by Council and College, though distinguished practitioners could be directly elected FRCP (Glasg.) without passing the MRCP on the basis of their published work or notable service. The FRCS (Glasg.) remained unchanged. The other changes that Shaw and Goodall incorporated into the order made Glasgow's the most democratic College, with all Fellows and Members enjoying equal voting rights in the election of Council and office bearers. The power to levy annual subscriptions was also formally incorporated into the order. The Council was henceforward to consist of the principal office bearers (who were elected annually and were members of Council *ex officio*), plus a further eight, six of whom were elected by the Fellows and Members, one by the Licentiates, one by the Licentiates in Dental Surgery, plus the College's representative on the GMC.[78]

As the GPGMB's submission to the Nuffield stated, as part of its upgrading into Glasgow's centre of postgraduate operations during 1962–63 the new College acquired a well-appointed new reading room by the beneficence of Sir Hugh Fraser (after 1964 Lord Fraser of Allander), who had already proffered crucial help in the name change negotiations. Fraser, 'undoubtedly one of the most spectacular figures in the British retail industry',[79] had assumed control of the family firm of Fraser and sons (House of Fraser) in 1927. As well as expanding the department store empire, he had also diversified into tourism, and later, in 1964, helped to see off a hostile take-over bid for the *Glasgow Herald* from the Thomson Organisation. He was a fierce patriot and generous philanthropist, after 1960 through the Hugh Fraser Foundation, which was endowed with £2,500,000. He had already been awarded the Faculty's Honorary Fellowship in 1959 for his support of medical education and research. Wright had met Hugh Fraser when he was in his late twenties, and Fraser in his early twenties. As Wright recalled in 1985:

He came to consult me in Bath Street [where Wright had his extensive private consulting practice where he treated many of the local business community], sent by a mutual friend and was suffering from aortic disease, post rheumatic following a chorea and we became close friends almost from that time until the day he died. When I was President of the College, he was very keen to do something for the College and I suggested the Library and he put up the finances

[78] 'Draft Provisional Order to Change the Name of the RFPSG to RCPSG', March 1962, RCPSG, 1/1/1/21.

[79] Quotation and following biographical details from A. Slaven and S. Checkland, *Dictionary of Business Biography, 1860–1960*, ii, *Processing, Distribution, Services* (Aberdeen, 1990), pp. 362–63.

for this. He also prepared the President's room, furnished it and helped in the furnishing of what became known as the Princess Alexandra Room.[80]

Fraser's wife was also involved, organising the decoration of the Princess Alexandra, Ladies' and President's Rooms. In all, the Hugh Fraser Foundation contributed £22,500 to the development of the Fraser Reading Room, which Sir Hugh opened on 1 May 1963.

Further internal administrative developments followed in 1963. R. B. Wright, the Honorary Treasurer, argued that there was now too much business to get through at Council meetings, and that, consequently, important issues often received only scant attention. To remedy this he suggested holding pre-meetings of the office bearers to dispose of routine matters before the full Council meeting and also the creation of two new advisory committees, one in Medicine and one in Surgery. The Council would refer relevant matters to these committees for a fuller investigation, and their reports, presented before the relevant Council meeting, would, 'enable informed and enlightened discussion to take place'.[81] The Medical and Surgical Advisory Committees (MAC, SAC) would have a stable membership drawn from senior College men, including those with wide experience of problems of medical education outwith Glasgow, and of the problems of junior hospital staff. At the 16 April Council meeting Wright's idea was implemented, with his suggested membership (each committee to have two Councillors and five from the Fellows or Members – including one from the junior staff group, one from outwith Glasgow, and one past-Councillor). The committees were to be appointed annually by each new Council at the November AGM, and business was to be referred for report either by the Council or the President. The first personnel were appointed to the new MAC and SAC at the 13 May Council meeting.[82] In November Dr Thomas Semple was appointed as the first Honorary Registrar for the Membership Examination, since the number of candidates had risen dramatically since the introduction of a Glasgow MRCP, and the College Registrar, Mr Willox, could no longer cope on his own with the organisation.[83]

Under pressure from the Royal Medico-Psychological Association, which

[80] Joseph Wright, transcript of interview with Peter MacKenzie, RCPSG, 8 February 1985, RCPSG, 18/20.

[81] 'Memorandum from Mr R. B. Wright on Method of Conducting the Affairs of the Council', April 1963, RCPSG, 1/1/4/21.

[82] MAC: Professor Thomas Anderson (Chairman); and Drs T. D. V. Lawrie, T. J. Thomson, John Stevenson, J. B. Barr, Peter MacFarlane and Thomas Semple. SAC: Mrs R. B. Wright (Chairman), Kenneth Fraser, D. C. Miln, John Todd, Alex Lyall, W. Ian Gordon and John Hutchison. *Council Minutes*, 13 May 1962, RCPSG, 1/1/3/14.

[83] Ibid., 4 November 1963.

was threatening to break away and turn itself into a separate Royal College with a separate Membership, the RCPL announced in May 1964 that its MRCP was to be radically altered. As a response to rising numbers of candidates, there was henceforward to be pre-selection of candidates via a new Part 1 screening examination to be composed entirely of multiple choice questions (MCQs). Part 2 was to be a general medical examination, but with question choice, and Part 3 introduced a new limited specialist element (Internal Medicine, Psychiatry, or Paediatrics). As Rosemary Stevens has noted, this, the first major change to the examinations of the London Physicians since the MRCP had been introduced just over one hundred years before (1861), was, 'a landmark in the College's history, as well as an apparent attempt to restrain certain groups from separatist action'.[84] In January 1965 the RCSEng was to alter its conditions of entry for the FRCS to four years (instead of three) of clinical experience in approved posts.[85] The Glasgow College was aware of the imminence of these moves early in 1964. In January Illingworth, now President, wrote an open letter circulated to all Fellows and Members seeking support for his plan to extend Glasgow's training requirements in its higher examinations to three years for Physicians and four years for Surgeons. He argued that,

> possession of our Membership or Fellowship should provide evidence that the holder has acquired an adequate amount of clinical experience and has had the opportunity to acquire proficiency in technical skill [and] that overseas candidates should take back something more of our way of life and approach to medical problems than can be gained by brief attendances at a coach class.[86]

Illingworth was clearly concerned that while the higher examinations of the Colleges were not examinations denoting full specialist status, they should be seen as indications of potential to become a specialist after further training. This meant that they should be sound tests of acquired general clinical skill. However, if this was to take three or four years to build up, a tension began to surface between the emerging concepts of the period of general training after which the MRCP/FRCS would be taken, and the following period of specialist training. These concepts had not fully emerged yet, and indeed, this tension was part *of* that emergence, which was ongoing throughout this period before the Todd Report of 1968.

This letter led to two lively Council debates. In the debate on medicine, Professors Alstead and McGirr led a successful opposition to Illingworth's

---

[84] Stevens, *Medical Practice*, p. 346.

[85] Ibid., p. 351.

[86] Letter from Sir Charles Illingworth, PRCPSG, to all Fellows and Members, January 1964, RCPSG, 1/1/4/22.

plan, and the conditions of entry to the Membership remained unchanged at three years' hospital experience. Perhaps they were mindful of the very tension described above. In the surgical debate Arthur Jacobs opposed, but R. B. Wright, J. H. Wright and Shaw ensured that here the Council's plan was implemented. The training time was changed from three to four years, with a detailed breakdown of how these years should be spent the same as the initial 1962 proposal, but with an extra year of specialist experience added at the end. These changes were to come in from 1 January 1965.

In September 1964 it was the turn of the content of the Surgical Fellowship to come under renewed scrutiny. The College SAC, which had been reviewing this, now advised that the 1962 suggested new regulations now be enforced, and that the option of an entirely specialist Surgical Fellowship, available since 1949, now be withdrawn. A limited specialist element was retained but now, once again, all candidates had to take part one of the examination in the general Principles of Surgery, before taking part two, in which they could choose between being examined on one of the four available specialties (Practice of Surgery – now considered a generalist specialty, Obstetrics and Gynaecology, Ophthalmology, or Otolaryngology).

In his end of year report President Illingworth was upbeat about changes in Surgery and it seemed that moves towards a common Fellowship might come there first:

> In the sphere of external relations, the status of our diplomas of Fellowship and Membership must always be a matter of first concern. Shortly after our change of name we agreed that, now we are to confer diplomas of MRCP and FRCS, it is more than ever important that such diplomas should be demonstrably equal to those of other Colleges. We later agreed in principle that possession of one of our diplomas should provide evidence that the holder has acquired adequate clinical experience and has had opportunity to acquire proficiency in technical skill. We have now implemented these recommendations in respect of our surgical Fellowship. The requirement imposed upon candidates to have undergone a four year course of postgraduate experience (including the pre-registration year) and that it should follow a definite though flexible pattern will ensure a proper postgraduate training. The examinations themselves are, of course, of a standard at least equal to those of other surgical Colleges. Our equality of status is now fully established. Our next step should be to consider, on equal terms with our sister Colleges, the possibility of unifying the surgical Fellowships so as to discourage the time-wasting and soul-destroying practice, which many postgraduates still find necessary, of taking multiple diplomas. We work on terms of complete reciprocity in the Primary examination and it should be possible to achieve something similar in relation to the Final Examination.[87]

[87] 'President's Letter', 2 November 1964 *AGM Billet*, RCPSG, 1/1/5.

As an eminent surgical President, Illingworth was no doubt particularly pleased with this seemingly rapid progress in his own branch of medicine. Further changes were to come in Surgery. In 1968 the Council stated its intention of coming into line with Edinburgh and Dublin's practice in the Final FRCS examination by replacing Principles of Surgery with a paper on Operative Surgery and Surgical Pathology (including questions on topographical anatomy) with orals in both.[88] In May 1969 the content of the Surgical Fellowship was further discussed. The FRCS was now altered to have a general Part 1 of written, oral and clinical papers in the Practice of Surgery. Part 2 was specialist options, but the first (Surgical Pathology and Operative Surgery) was a generalist subject geared towards those intending to specialise in General Surgery. The other options remained unchanged.

As early as December 1968, the British Surgical colleges had together agreed that MCQs should replace the existing six compulsory written questions in the Primary examination of all colleges by May 1969.[89] This seemed like another move towards common examinations, in this case from reciprocity between Primaries towards a Common Primary. The RCSEng had been the first college to institute a Primary in 1867 based on the pre-clinical subjects of Anatomy and Physiology. The RCSI introduced a similar examination in 1888, and the two Scottish bodies followed suit rather later, in 1949. Reciprocity between Primaries was achieved in 1951, and in 1960 all four bodies introduced a tripartite examination in Anatomy, Physiology and Pathology. Since 1955 representatives from the four had met annually to review examination issues. In 1967 the SAC of the Glasgow College concluded that the existing content of the examination bore little relevance to the actual work of young hospital doctors in the first two postgraduate years. They recommended that the examination become a test of the clinical application of the basic medical sciences, rather than remain as a reexamination of the more theoretical aspects of the basic sciences, which the undergraduate curriculum and examinations now covered in depth. They also wished to move away from essays and towards MCQs as a more appropriate, 'objective', way of examining this more standardised form of knowledge. This impulse was itself an indication of the high degree of scientisation of this basic level of clinical skill. At the international Joint Conference of Surgical Colleges held in Melbourne in May 1968, Glasgow was authorised to pioneer the unilateral introduction of an experimental MCQ-based Primary as a pilot study before the other colleges adopted it. Just before the first such

[88] *Council Minutes*, 12 November 1968, RCPSG, 1/1/3/17.
[89] Ibid., 10 December 1968.

examination was held on 19 May 1969,[90] the College issued an explanatory announcement to the medical press. This statement illustrates that issues of specialisation and the synchronisation of examinations with the hospital training ladder were uppermost in College minds. The two main criticisms levelled at the existing examination were that it was:

> unduly punitive to the young man in training and that it demands a great deal of reading, much of it revision and much of it not of particular application to the specialty he has chosen to follow. This criticism is especially valid in regard to topographical anatomy. Secondly, the examination is not entirely relevant to the clinical situation in which the candidate is working. Only the exceptional candidate can take the Primary in his stride. The majority have perforce to sacrifice time reading for this examination that might be better spent in the wards or laboratory.[91]

The MCQ format would also be useful as a preliminary screening exam to whittle down the increasing number of candidates. The statement went on to argue that *all* hospital doctors, physicians or surgeons, needed 'a thorough background knowledge of cell and organ structure and function and of the pathological variations from the normal'. It suggested that in future therefore all intending specialists should take a common Primary examination in 'Applied Human Biology'. It expressed the hope that: 'Such an examination could replace the Primary FRCS examination, the Part 1 MRCP examination, and perhaps also Part 1 of the MRCOG examination'.[92] This is a telling indication of just how concerned with standardisation of examinations the Colleges were in the 1960s. Also there is a hint here that standardised examinations, perhaps run by national umbrella organisations representing all Colleges (and perhaps the specialist associations), were a means of reintegrating specialists into a unified examination structure. This joint approach at a supra-collegiate level was to be tried again very successfully later with the inter-collegiate specialty examination in the Surgical specialties.

MCQs were sensitively introduced into the Primary in four phases. In each phase an increasing amount of the exam was in MCQ format, until in Phase IV, from October 1971 to January 1972, only those candidates with a

[90] See Stuart Young and Gordon Gillespie, 'Experience with the Multiple-Choice Paper in the Primary Fellowship Examination in Glasgow', *British Journal of Medical Education*, 6 (1972), pp. 44–52. Interestingly, Illingworth had been a pioneer of MCQs in the degree examination for Surgery at Glasgow University, see his, 'The "Multiple Choice" or Objective Examination: A Controlled Trial', *Lancet*, 14 December 1963.

[91] 'The Primary FRCS Examination: A Fresh Look', *BMJ* (1969), 1, p. 502.

[92] Ibid.

'reasonable' mark would take an oral, the rest would have passed or failed on the MCQs alone.[93]

By 1970, however, this tide of progress in the Surgical examinations seemed to have run out. The then President, R. B. Wright, concluded after meeting with the Presidents of the other Surgical colleges in April that, 'It was clear that the time had not yet come for acceptance of a common primary because of the unsettled problems of definition of training programmes'.[94] At the same time, after discussing a memorandum on FRCS reciprocity by Professor Andrew Watt Kay, Council added to the gloom by noting that, 'it became increasingly clear that Commonwealth reciprocity could not be completely achieved so long as the British Colleges gave an FRCS Diploma at the end of basic training, while the Australasian, South African and Canadian Colleges preferred to give their Fellowship at the end of full specialist training'.[95] The Surgeons had fallen away from their earlier cooperative strategy. There had been a tension about when the Final examination should be taken, and what it should denote, throughout the 1960s. By 1970 a lack of agreement of what the Fellowship should indicate about the successful candidate (capability to practise as a specialist or merely potential after a further period of training), as well as difficult financial arrangements, had derailed the standardisation process. Reciprocity of Primaries was itself lost later on. However, during the 1990s the issue of the standardisation of surgical examinations was again to be addressed by the Royal Colleges.

In fact, progress towards a common vision of a common examination structure, with a commonly agreed function and timing on the training ladder, was achieved more rapidly in medicine. It had not seemed this way back in 1964 when *lack* of progress in reforming the Glasgow MRCP had moved Gavin Shaw as Honorary Secretary to warn College after the Surgical Fellowship had been changed again that year:

Proposals of a similar nature in regard to the Membership did not find favour with the College and the whole question of the status, function and form of the Membership examination of the College will require to be most closely reviewed in the months to come. The examination must be considered in its relationship to the training of a consultant physician and also in the light of changes occurring in the examinations of other Colleges; moreover, the needs of specialists will

[93] See Stuart Young, 'Addendum to the Glasgow College's Memorandum: Melbourne 1968', April 1972. I am grateful to the author for sight of this from among his working papers.
[94] 'Report by President on Meeting of Joint Conference of Surgical Colleges', 1 April 1970, with *Council Minutes*, 14 April 1970.
[95] From discussion at Council Meeting on Professor Andrew Watt Kay's memorandum on 'Reciprocity of FRCS Diplomas', ibid.

have to be taken into account in these times of changing medical fashions and practice.[96]

It was the standardisation of the UK Physicians colleges' MRCP examinations that was to prove the success story of this period.

Meanwhile the Glasgow College's Dentists had also been influenced by the mood of specialisation and were considering the best means of improving the status of their examinations. In February the Convenor of the Dental Council, Professor T. C. White,[97] was pressing for the Higher Dental Diploma (HDD) to become a Fellowship in Dental Surgery: FDS, RCPS Glasg.[98] The Dentists argued that:

> It is becoming clear that the Higher Dental Diploma of this College is not regarded, in other parts of the country, as being the equivalent of a Fellowship in Dental Surgery. Such Fellowships are granted by the RCSEng, by the RCSEd and the RCSI. Our Higher Dental Diplomates find themselves competing for posts on unfortunate terms with the holders of Dental Fellowships. The Directors of the Dental Branches of the Army and the RAF have stated that an FDS is a necessary qualification for specialisation within the service, the HDD not being regarded as adequate. Recently it has been ascertained that possession of the HDD may not entitle the holder to be included in the 'Colonial List of Higher Dental Qualifications'. *It seems that those circumstances which have caused the Royal Faculty to change its name also apply in Dentistry.*[99]

The plans for the extension of the University Dental School would mean a large increase in the number of postgraduate students. The establishment of a new dental Fellowship (with hospital-based training schemes) at this moment, would therefore re-cement the relationship between the College and dental students, which had broken down when the BDS replaced the LDS as the basic qualification in 1948. Currently, 'Holders of Diplomas of this

[96] 'Report by the Honorary Secretary on Items of Interest from the Minutes of the Council, 1963–64', 2 November 1964, *AGM Billet*, RCPSG, 1/1/5.

[97] Thomas C. White (1911–80). Born in Falkirk and the son of a dental supplier, White took the TQ and the LDS (1935). After beginning an orthodontic specialist practice, he returned to Dental Hospital and School at Garnethill and began to develop the Orthodontic Department. He remained at the School when it was taken over by the University, becoming a consultant to the WRHB, and Professor of Orthodontics in 1963. In 1964 he succeeded Professor Aitchison as University Dental Dean and presided over the building of the large extension to the Dental Hospital, which was opened by HRH the Duchess of Kent in 1970 and provided an improved clinical, teaching and research environment, used by the School's expanded undergraduate intake of approximately seventy per year. See Ian Melville, 'T. C. White', *College Bulletin*, 18 (May 1989), pp. 33–34.

[98] *Council Minutes*, 18 February 1964.

[99] 'RCPSG Proposed Fellowship in Dental Surgery Explanatory Leaflet', 6 March 1964, RCPSG 1/1/4/22. My emphasis.

College have no sense of affiliation to the College: which they might otherwise regard as their *Alma Mater*.[100] White's initial plan had been to follow the RCS's route and create a Faculty of Dentistry within the College to run dental examinations and handle all other dental affairs.[101] The Faculty would supersede the Dental Council and would elect a Dean. At its 18 February meeting, the Council approved of the plan for the FDS, but balked at the idea of creating separate Faculties within the College. Perhaps the prospect of an independent College of Pathologists, which was to be announced in April, had formed a determination to keep breakaway movements in check in Glasgow, if at all possible. However, a Dental Faculty was later established in 1990.

During the rest of 1964 the College gained annual library grants of £850 for textbooks and evening opening from the WRHB under the same new SHHD funding that had this year taken over the funding of the clinical tutors. Previously these library grants had also come from the £25,000 Nuffield grant to Glasgow, but they were now made permanent WRHB grants.[102]

The College was also involved in discussions with the other Scottish Colleges, the universities and RHBs on the implementation of the Wright Report's recommendations on the ideal consultant numbers in different specialties. Mostly this meant pushing for increases in the consultant establishment so as to take some of the work off the shoulders of those in the so-called training grades.[103]

By the latter part of the year, however, the focus had returned to the Membership. In September, Shaw noted that the 310 candidates then proposing to sit the Membership at the next diet was 'the absolute limit to which the host examiners in the City of Glasgow could be stretched in regard to numbers'. Shaw suggested that the Council find some way of limiting the numbers presenting in future.[104] In early October the sub-committee on consultant physician's training of the standing joint committee of the three Scottish royal colleges published its report. The report embodied an emerging consensus within the profession in Scotland on the place and purpose of Membership examinations, later to be given its full expression in the Todd Report. The period of consultant training should, the sub-committee argued, be divided into 'basic general training' and subsequent 'specialist training'.

---

[100] Ibid. The final part of this sentence was removed from the final draft.

[101] The RCSEng has established Faculties of Dentistry and Anaesthesiology to handle the new exams of FFDS and FFA in 1947 and 1948 respectively. Stevens, *Medical Practice*, p. 113.

[102] *Council Minutes*, 21 April 1964.

[103] Ibid. J. H. Wright chaired the SHHD's Committee which reported on *Medical Staffing Structure in Scottish Hospitals* (Edinburgh, 1964).

[104] *Council Minutes*, 15 September 1964.

The former would comprise the pre-registration and post-registration year, which should be spent in acquiring a broad basis in 'general hospital training'. Trainee consultants would then spend six to eight years in specialist training in Registrar and Senior Registrar posts. With this basic structure agreed it was then possible to define the Membership as a passport to higher specialist training and to thus synchronise both its timing and its content with the realities of hospital clinical work and the training ladder:

> the aspirant to consultant status should undergo an initial proving period in which his potential and right to train will be assessed, and a subsequent longer period of intensive specialist training. The gaining of a Membership of a College of Physicians is seen as one of the important demonstrations of *potential* during the proving period and not as a diploma of specialist ability.[105]

Armed with this consensus, the Colleges could begin to move forward towards standardisation.

During 1965 all the Colleges of Physicians were experiencing problems coping with the demand for their Memberships. The RCPL introduced a MCQ Part 1 as a screening examination to weed out unsuitable candidates and thus reduce the pressure on Part 2 from May 1964. The Edinburgh College, which had been forced to utilise extra patients in Dundee and Glasgow hospitals for its clinicals, followed suit in 1965 and introduced a written Part 1 screening examination of its own to operate from 1 January 1966. There were 727 candidates for the Glasgow Membership in 1965, up again from 638 the previous year.[106] As W. G. Manderson noted, 'The examination had to be held in many hospitals in and near Glasgow and the clinical examination of so many candidates placed an intolerable burden on all concerned with its organisation'.[107] To remedy this the College considered the idea of an elementary clinical examination as a preliminary screening test. The Visitor, James Holmes Hutchison, suggested coopting Sir Edward Wayne, who had succeeded McNee as Professor of Medicine in the University Department of Medicine at the GWI in 1953, and Dr Eric Oastler, both of whom, 'had close connections with the London College', on to the Glasgow College's MAC. Hutchison was, however, gloomy about the prospects for cooperation with London over the Membership, but thought it might prove easier with Edinburgh. He was concerned that the MAC's current plan for the Membership favoured specialists less than the Edinburgh and London plans, but Thomas Anderson stressed in Council that the MAC's report had

[105] 'Standing Joint Committee: Sub-Committee on Training of Consultant Physicians', RCPSG, 1/1/4/22. Original emphasis.

[106] 'Honorary Secretary's Report', *AGM Billet*, 1 November 1965, RCPSG, 1/1/5.

[107] Manderson et al., 'Evolution of an Examination', p. 101.

been unanimously in favour of a more generalised examination.[108] By October, however, the MAC had accepted the need to make the revised Membership appeal to specialists and had tentatively agreed that it should contain a limited specialist option in Dermatology, Paediatrics, Psychiatry, and Social Medicine and Epidemiology.[109] The new regulations for the Membership were discussed from the end of 1965, and put to College and accepted in January 1966. They introduced a new four stage examination. Part 1, taken one month before Part 2, consisted of twenty compulsory questions on the Practice of Medicine including the related medical sciences and Therapeutics. Part 2 was now a two-stage written examination. The first paper contained four compulsory questions and was a wide-ranging general medical paper, including clinical applications of the basic medical sciences. The second paper had two compulsory questions on the Practice of Medicine, and then had ten pairs of questions in the Practice of Medicine (for specialists in general medicine), Dermatology, Paediatrics, Psychiatry, and Social Medicine and Epidemiology. The candidate preselected his special subject for this part of the examination, then answered those two questions only. The examination was completed with a general clinical (including ancillary methods of diagnosis and treatment), and two fifteen minute orals (one in Practice of Medicine and one in the special subject).[110] The first Glasgow screening examination was held in September 1966, when there was a 33 per cent pass rate. This reduction in the number of candidates for Part 2 was by then even more desperately needed, as the numbers presenting for the Membership had dramatically increased again during 1966 to 903. This was the highest ever figure, but was partly explained by the fact that the June 1966 diet was the last at which the examination could be taken at a single sitting.[111]

While negotiations with the two other Colleges of Physicians continued, the Glasgow College was also considering its submission to the Royal Commission on Medical Education. Appointed on 29 June 1965 and meeting first in September, the Commission, and its 1968 Report, was known after its chairman, Lord Todd. It contained very eminent members of the profession (including three from Glasgow, Professors Andrew Watt Kay, G. M. Wilson and Charles Fleming). It had been given a wide remit to review both undergraduate and postgraduate medical education and to recommend the principles on which future developments should be made. The committee drew on the debate already ongoing within the profession about how the

[108] *Council Minutes*, 17 February 1965, RCPSG, 1/1/3/15.
[109] Ibid., 19 October 1965.
[110] 'Report of the MAC on the Content of the Membership Examination', December 1965, *Council Minutes*, 21 December 1965.
[111] *AGM Billet*, 7 November 1966, RCPSG, 1/1/5.

education of doctors should respond to the growth of specialised medical knowledge, and how it should be organised within the NHS, which has been traced here.[112]

As part of the preparation of its submission, the Glasgow College had begun a process of consultation with the local hospital staffs, most of whom were Fellows or Members. In August 1965 R. B. Wright had written to all of the local district hospitals seeking a description of the postgraduate training currently on offer, and suggestions for improvements.[113] In May 1966 the College extended a general invitation to all consultant surgeons in the district hospitals in the Western Region to meet with the SAC to discuss the Todd submission.[114] Just how closely the consensus later expressed by the Todd Report was based on the ongoing debate within the profession is shown by the fact that in late 1966 a London meeting of all the Royal Colleges discussed the possible future institution of a Specialist Register to be held by the GMC. The meeting recommended that all Memberships should be considered as equally qualifying the holder for NHS posts, and that training posts should be organised by regional postgraduate committees under a national coordinating body.[115] The submissions to the Todd Commission from Glasgow College and from the RCPEd were likewise very similar. Both supported a two year pre-registration period, the institution of rotation programmes for senior hospital staff, and the need to relate more closely the numbers in each training grade to later opportunities in the hospital service. There was agreement, too, on the basic timing and significance of higher qualifications. These should be early in the hospital career, after a period of general training, and should facilitate the beginning of further specialist training, which would be laid down by the Colleges for different specialties.[116]

The growing consensus between the Colleges on postgraduate education found further expression in the Scottish Postgraduate Medical Association's 'Report of a Working Party on Postgraduate Medical Education within the National Health Service in Scotland', which was chaired by J. H. Wright and published on 7 April 1967. The report, which was criticised and disowned by the SPMA, probably because it recommended the establishment of another national coordinating body for postgraduate education, echoed the post-

---

[112] *Royal Commission on Medical Education* (Todd Report), Cmnd 3569 (London, 1968), pp. 8–22.

[113] See the replies to this circular letter in RCPSG, 1/1/11/1 (f).

[114] *Council Minutes*, 17 May 1966, RCPSG, 1/1/3/16.

[115] Ibid., 22 November 1966.

[116] 'RCPSG Memorandum to the Royal Commission on Medical Education', May 1966, *College Minutes* (uncatalogued), 2 May 1966. For details of the Edinburgh submission see Craig, *History of the RCPEd*, pp. 838–50.

Christ Church plans which had already been largely implemented in Glasgow. The working party's report further stressed the still disputed point that all Colleges must now accept that their Memberships and Fellowships should be taken early in hospital training after a period of general and before a period of specialist training. The working party's calls for a new national coordinating body were answered when the Central Committee on Post-graduate Medical Education was formed in early 1967, after discussions between all the Colleges, the Committee of Vice-Chancellors and Principals (CVCP) and the Ministry of Health in late 1966. The new committee was soon extended to include Scottish representation as the CCPME (GB). It functioned largely as an advisory body, acting as a clearing-house on post-graduate schemes all over the country, and advising the regional postgraduate committees. It formed an important national interface between the Royal Colleges, the Universities and the Health Departments, offering the government the possibility of a single authoritative source for advice on postgraduate medical education.[117]

Meanwhile there had been further progress on the Membership. A joint meeting of all the Colleges of Physicians in London on 26 May 1967 had agreed that a *common* Part 1 should be instituted as early as possible. It should be an all MCQ examination marked by computer, controlled by a conjoint board and taken two years after graduation.[118] In his end of year letter to Fellows and Members, the President, Holmes Hutchison, earnestly hoped that these developments might mark a new consensus on the timing and purpose of collegiate higher examinations:

> The successful completion of the period of basic vocational training for the young doctor intent on a career in hospital medicine in this country has traditionally been marked by the acquisition of the MRCP/FRCS or equivalent diplomas. The tendency in recent years for men to obtain multiple diplomas has been increasingly and rightly deplored as a waste of valuable time and potential. The remedy lies with the Colleges, and our College is at present engaged in discussions with the Royal Colleges of Physicians of London and of Edinburgh towards the introduction of a Common Part 1 to our Membership examinations. While this will probably be achieved in the near future, the final solution to the problem of 'multiple diplomatosis' must ultimately be found in some type of Common Membership examination. There is, of course, already reciprocity in the Primary FRCS examinations of the surgical Colleges, but here, too, some form of common final examination seems to me a possibility. Indeed, it is possible that the whole content of our senior diploma examinations will come under scrutiny as a result

---

[117] John Lister, *Postgraduate Medical Education*, the Rock Carling Fellowship 1993 (London, 1993), pp. 27–28.
[118] *Council Minutes*, 20 June 1967.

of the present discussions between the Colleges ... never before have the Royal Colleges engaged in such close cooperation, and the implications may be far-reaching.[119]

The new preliminary part to the Membership was approved by College in April 1968 and the first Common Membership Part 1 examination was held in the College in October. The Part 2 examinations remained, for the present, different in each of the Colleges of Physicians.

On other fronts, 1967 finally saw the establishment of a Dental Council to advise the President and Council on dental education and training and on the FDS examination. The President, Visitor and the Dental Representative on the College Council were members, four members were to be elected by the new Fellows in Dental Surgery, one by the Council, and one by the College. The Dental Council itself was to have representatives on the GDC, the GPGMB and the Scottish Committee for Hospital Dental Services. The new Council was soon organising courses in Advanced Dentistry with the GPGMB for candidates for the FDS. The FDS had been accepted as an additional registrable qualification in November 1966, and the first FDS examination was held in March 1967. By the end of 1967 there were forty-nine dental Fellows, though this included a number who had been elected as founder Fellows on the basis of their existing status and experience. In 1969 a Joint Committee on Higher Training in Dentistry was formed from the Surgical Colleges, with meetings paralleling those of the Joint Committee on Higher Surgical Training (JCHST) which was also formed that year. The College Dental Council sent two representatives, being also represented on the Specialist Advisory Committee on Dental Surgery and Orthodontics.[120]

Closer to home, 1966 had also seen the retirement of the College's long-serving General Registrar, Mr Willox, who before the appointment of individual Honorary Registrars for the Membership and Fellowship had undertaken the entire organisation of College examinations, and much else besides. It is a mark of the importance of his contribution that he received a gratuity of £1250 for long and faithful service, a public notice of his retirement being inserted in the *Glasgow Herald*. He was succeeded on 2 May 1966 by John W. Robb.[121] In 1967 the College further developed its administrative structure by appointing Stuart Young to be the first Honorary Registrar for the Surgical Examinations. With the parallel post of Honorary Registrar for the Membership examination having been established in 1963,

[119] *AGM Billet*, 6 November 1967, RCPSG, 1/5/5.

[120] See J. A. Russell, 'Dental Council', *College Bulletin*, 1 (November 1971), p. 9.

[121] *Council Minutes*, 25 January 1966.

the work of the College's General Registrar was now reduced to a more manageable level.

Once a Common Part 1 had been achieved for the Membership, attention moved to the reform of the heart of the examination, Part 2. The College Meeting of 5 February 1968 sanctioned the integration of Part 2s, and in July the President, Holmes Hutchison, and Professor Edward McGirr attended a joint meeting in London of all the Colleges of Physicians which addressed this issue. Holmes Hutchison's report of this meeting shows that while the Colleges were moving towards the idea of the higher qualifications as an index of potential taken early in the career, there was still much debate on this issue. However, there was also the will to standardise:

> After prolonged and detailed discussion it was agreed that the ultimate aim must be a Common examination in all its parts. It was accepted also that the content of such a Common Part 2 examination should be subjected to complete re-appraisal. It might be, for example, that the Part 2 examination would become more directly related to the training of specialists in certain broad areas of medicine, for example internal medicine, paediatrics, psychiatry. It is obvious that such far reaching changes cannot and should not be introduced quickly. At the same time the need to deter young men from sitting multiple Memberships now is strongly acknowledged.[122]

Thus, as soon as the Common Part 1 was introduced in October, all the Royal Colleges agreed to 'introduce *complete* reciprocity between their Memberships. This interim arrangement would exist until a Common Part 2 examination could be planned and instituted'.[123] Reciprocity would apply only to those first passing the new Common Part 1. A joint public announcement was to be made saying that the Colleges regarded each MRCP as of equal standing and also, therefore, deprecating the need to pass more than one. To underline this the existing scheme for interchange of examiners between the Colleges would be increased. When agreed, the new Common Part 2 would be administered by a common UK examining board, and be taken four years after graduation ('and often earlier') by successful Common Part 1 candidates.[124] In June the President confirmed in a letter to Members and Fellows that the joint public statement had indeed been issued. Confirming the *laissez-faire* nature of British postgraduate qualifications, Hutchison doubted that this statement would necessarily have any immediate

---

[122] 'RCPSG: Memorandum by the President on the Progress of Further Discussions of the MRCP Examination by the Committee of the Three Royal Colleges', RCPSG, 1/1/1/24.
[123] Ibid. Original emphasis.
[124] Ibid.

impact on the decisions of hospital appointments committees. Reconfirming his own support for a Common Part 2 he added that:

> Indeed, it is my own opinion that none of the MRCP examinations, as at present constituted, is really relevant to the training of the modern physician, and that the time is opportune to make a fresh appraisal of the whole content of the Part 2 MRCP examination. It is, of course, one thing to express such opinions, quite another to assume that the Fellows of our sister Colleges will hold similar ones. The only forecast which I would care to make about our future discussions is that they will be difficult and complicated.[125]

What Hutchison had perhaps not expected was that there would also be difficulties about the recasting of the MRCP from within his own College. These differences of opinion were to emerge after the publication of the Todd Report which, while partly expressing the emerging consensus on postgraduate education, also attacked the relevance, and the very existence, of the Colleges' higher qualifications as part of that educational structure.

The Todd Report expressed the consensus that had emerged within the profession on the basic structure and aims of medical education after the first degree, and how this should be related to the NHS staffing structure. Changes in medical knowledge with the rise of specialisation meant that it was now necessary for hospital consultants to be regulated. The Medical Act of 1858 had done this for undergraduate qualifications, but now the postgraduate period needed to be subject to a measure of standardisation and professional scrutiny. After the pre-registration appointment, which would continue to be organised by the universities, those intending to become hospital consultants/specialists would spend three years in 'General Professional Training' (GPT).[126] This period would be common to all doctors and would include basic clinical and research training slanted towards the general field in which the doctor hoped to make his career. Doctors would progress through a planned series of six to twelve month appointments, differing with each specialty. Monitored by a new national Central Council for Postgraduate Medical Education, the regional postgraduate committees would organise GPT, and like the membership of those bodies, it should reflect a blend of academic and clinical conceptions of medicine. While GPT would be 'vocational rather than academic', nevertheless, 'emphasis should be placed throughout on applied scientific methodology'.[127] Todd supported the development of the Nuffield/Goodenough agenda of increased involvement for the regional university in postgraduate education. The universities would

---

[125] 'Letter from the President', June 1968, RCPSG, 44.
[126] *Todd Report*, p. 42. paragraph 59.
[127] Ibid., p. 49, paragraph 78.

thus now extend their influence into GPT. They would provide, 'teaching skills, facilities and – perhaps most important – a critical appraisal of standards which was independent both of professional practice and National Health Service requirements'.[128] In short, GPT was to be academicised, that is based on a more academic model of medicine as specialised and scientific.[129] The University Postgraduate Dean was to be the key figure, and in England teaching hospitals were to be included in the RHBs to achieve the same kind of overall central control and efficient use of regional training resources that had been achieved in Scotland. However, the Royal Colleges were also to have a role in GPT through their membership of the regional postgraduate bodies and through the clinical tutors in the regional hospitals.

After GPT would come 'Further Professional Training', designed to prepare specialists for consultant appointments. In this stage Todd envisaged a greater role for the Colleges in preparing detailed training and education requirements in the various specialties. After the successful completion of specialist training, doctors would be admitted to a specialist or vocational register, ultimately to be held by the GMC, which would indicate competence to exercise a substantial measure of independent clinical judgement.

The problem for the Royal Colleges came with the way that Todd envisaged the passage from GPT to specialist training being marked. Todd criticised the existing system of assessment at this stage – the higher qualifications of the Royal Colleges – on a number of counts. First they,

> vary widely in their design and intended significance. Those of the older-established bodies are designed primarily as a test of potential ability, aimed at selecting the doctors best suited to become consultants in due course rather than assessing what the candidates have already achieved in postgraduate training.[130]

The examinations of the older Colleges were taken on average five or six years after graduation, and even those of the newer Colleges were not related sufficiently to existing arrangements for postgraduate training in the hospitals. In addition to this 'inappropriate timing and nature', the present assessments were also 'uncoordinated', despite some recent progress towards common examinations. Todd argued that the early postgraduate phase should now cease to be dominated by preparation for formal examinations:

> One of the principles which we firmly hold is that in the assessment of general professional training there is no place for a single major 'pass' or 'fail' examination. Trainees should be assessed on a progressive basis throughout the three year

[128] Ibid., p. 79, paragraph 179.
[129] The Report later supported the Goodenough plan for putting all the teaching in all clinical departments under an *academic* head. See p. 213, paragraph 517.
[130] Ibid., p. 51, paragraph 87.

period, so as to relate the assessment closely to the trainee's particular experiences, to avoid some of the unreliability inherent in a once-for-all review covering the whole of the training, and to provide a basis for a reappraisal of progress and plans at intervals during the training period ... The final assessment should include a review of the interim assessments.[131]

The successful completion of GPT should rather be marked by the issue of a certificate of completion, which would indicate the completion of a planned series of training appointments in approved posts, with continuous assessment of the trainee's development by the regional postgraduate committee. The certificate should itself be regarded as evidence of eligibility for membership of the appropriate college, which should be regarded as a 'routine next step', without examination.[132] Todd further recommended that all Colleges should grant 'Memberships' once the certificate of GPT had been gained. There should be further moves towards reciprocity or common exams in both Medicine and Surgery. National MRCP and MRCS distinctions should be the ultimate goal. The 'Fellowship' should be regarded as a higher distinction granted at the end of specialist training, again not by examination but rather automatically on admission to the vocational register.[133] Todd was thus attempting to locate the definition of the consultant firmly within the hospital/university structure, and to exclude the Royal Colleges because their examinations were not properly geared to this structure. Todd therefore threatened the very foundations of the new role in postgraduate education that the Royal Colleges had carved out for themselves since the war.

The report meant that the Colleges had quickly to determine the exact place and purpose of their higher examinations, in order to present an alternative postgraduate structure in which they retained a key role for their formal examinations. In July 1968 the report of the RCPSG Council subcommittee on Todd cautiously welcomed its attempt to plan postgraduate education but insisted on a continuing role for the Colleges:

The general pattern of postgraduate training proposed by the Commission ... would appear to constitute a distinct advance on the somewhat unplanned pattern of training which is common in the UK at the present time. We are none the less of the opinion that the final assessment of the final completion of general professional training should include an appropriate examination.[134]

---

[131] Ibid., p. 53, paragraph 93.
[132] Ibid., p. 54, paragraph 98.
[133] Ibid., p. 54, passim.
[134] 'Report of The Royal Commission on Medical Education: Report by a Sub-Committee of the Council of the RCPSG', July 1968, with *Council Minutes*, 16 July 1968, RCPSG, 1/1/3/17. The report had been requested by the Chief Medical Officers of the Scottish Home and Health Department, and the Ministry of Health.

While allowing that the examination should take account of the candidate's continually assessed previous experience, the sub-committee argued that it should remain an *examination*, and that the Royal Colleges were 'particularly suited' with their 'unrivalled experience' to continue to guard the gateway to higher specialist training. The sub-committee agreed that there should be no further examinations at the end of specialist training, but that the Royal Colleges should be responsible for accrediting successful completion of further training before the doctor was included on the vocational register. At the College meeting on 2 September there was a lively debate about proposed changes to the Membership examination, which reflected the need, in the immediate aftermath of the potentially damaging Todd proposals, to adhere to the joint position of the Royal Colleges of Physicians, and the tensions surrounding this position:

> There was varying opinion regarding the correct timing of specialist training. One school of thought was that candidates should pass the Membership Examination and then complete a necessary period of specialist training before being admitted to any specialist register. Another school of thought was that the Part 2 Examination should be the final step and that candidates should only be admitted to this examination after having undergone specialist training. On this latter point the President replied that ... the award of the MRCP Diploma under the present regulations could only indicate potential and that it had never been considered as being a qualification of specialty. To a suggestion by Mr A. B. Kerr that possibly candidates were admitted to the MRCP Examination at too early a stage after having graduated, the President replied that this view was not shared by the sister Colleges of Physicians. The feeling amongst the Physicians was that examinations should be disposed of as quickly as possible and so allow the young graduate to proceed to specialist training at an early stage in his career.[135]

The debate in the College is further illustrated by the suggestion put forward by Gavin Shaw at this meeting, then developed in a detailed memorandum. Shaw's response to the Todd proposals was to shift the Membership examination to the end of specialist training and to make it a specialist examination. Success in this examination would only then lead to the issue of a specialist certificate, not the other way around. Shaw also argued that the Colleges should run the continuous assessments of trainees, approve all posts and lay down the details of specialist training. Against Todd's support of a more active role for the universities and hospitals, he countered with a scheme for a more active College role, commenting that the Colleges, '*must* be in charge of the organisation of postgraduate training although the Universities and Regional Boards must be represented. The Colleges' diplomas must

---

[135] *College Minutes*, 2 September 1968, RCPSG, 1/1/1/24.

be the seal of *full* training, not potential'.[136] Shaw, who had been one of the main architects of the College's new role in postgraduate education, was not about to let that hard won role go without a fight. He stressed that there should be a new regional postgraduate board in Glasgow, one which was completely independent, and did not appear to be still essentially a university sub-committee. This last suggestion prefigured the establishment of the Western Regional Committee for Postgraduate Education in 1970.

However, as President Hutchison had indicated, this route was not the one preferred by the sister Colleges, and he was determined that Glasgow should keep to the joint position in the battle against the Todd proposals on higher qualifications. Glasgow's strength lay in unity with the other Colleges. This meant continuing the process of standardising the examinations. In Surgery, as noted above, there was some progress, but the real hope was for the Common MRCP, and this, as the other Colleges agreed, was going to be a general examination taken early in the career. The Todd Report acted as a spur to the progress of these negotiations. Perhaps standardisation of examinations would alleviate the Todd threat to the continued existence of the higher qualifications as passports to higher training. In any case, by the end of 1968 the Common Part 1 MRCP was up and running and negotiations were continuing in the Joint Standing Committee of the three Scottish royal colleges on reciprocity between the Part 2s.

The College Council was determined to keep the College in the mainstream of thinking on the place and content of the Membership examination so that Glasgow could be part of coordinated national developments. The points that the Council recommended for discussion at the joint meeting of the three Colleges of Physicians in June 1969 clearly demonstrate this. The Council argued that:

1. The Membership examination should be in General Medicine and of a standard relevant to the end of general professional training, after about four years.
2. It should not be considered to be a specialist examination; but it should continue to give the candidate credit for his having spent part of his general professional training in other disciplines.
3. The format should be such that it was relevant to a candidate's training and should set a standard which would select those who were suitable for specialist training.[137]

The College would now adhere to this line. Edinburgh too joined the com-

[136] Gavin Shaw, 'The Place of the Membership Examination' (1968), RCPSG, 44. Original emphasis.
[137] 'Memorandum on the Future of the Membership Examination', Council Memorandum in *Council Minutes*, 10 June 1969, RCPSG, 1/1/3/17.

mon position, in spite of concern over the threat to the retention of its specialist option in any future Common Part 2 and over the possible loss of revenue.[138] It was the successful joint response of the Colleges of Physicians to the threat posed to their examinations by Todd. In October full reciprocity between the Memberships was agreed between the Colleges. From now on candidates successful in the final examinations would become MRCP (UK). They would then apply for collegiate membership of the College of their choice, and this would be reflected in the title of their qualification, for example, MRCP (Glasg.), or MRCP (Ed.). The Common Part 2 was still being pursued, but now was only a matter of time. As W. G. Manderson, then the Honorary Registrar for the Membership Examination, commented in 1971:

> The net result of these changes and the discussions which have attended them has been clarification of the position of the Membership Examination in post-graduate training. The MRCP (UK) is now an examination directed towards assessing a candidate's potential for further training as a physician, this assessment normally taking place two to three years after qualification.[139]

The introduction of the Common Part 1 screening examination had done its job and had reduced the numbers going forward to take the Part 2. While this meant a more manageable workload, it also meant reduced fees. Consequently, the College ran a deficit of £1320 7s. 6d. at the end of 1969. To combat this, it was decided to begin to charge new Members and Fellows the annual subscription, whereas previously they had been exempt for the first five years. They now paid £6, and £10 *per annum* after five years. Those who had been consultants for five years now paid £20 p. a., and overseas Members and Fellows £6.

As noted, it was the threat from the Todd Report that had galvanised the Colleges into action to reform both the Medical and Surgical Examinations and to standardise them as examinations taken early in the consultant career. This was not the only way in which the stimulus of Todd shaped the template for the future of postgraduate medical education (and, within this, a new role for the Colleges) which is still being worked out in practice. In 1969 the Colleges established the Joint Committee on Higher Surgical Training (JCHST), with Specialist Advisory Committees (SACs) which included members from the specialist associations, to lay down formally the training requirements in the various surgical specialties during the period of further professional and specialist training. Medicine followed suit in 1970 with the establishment of the Joint Committee on Higher Medical Training (JCHMT)

---

[138] Craig, *History of the RCPEd*, pp. 773–74.
[139] W. G. Manderson, 'The Membership Examination', *College Bulletin*, 1, no. 1 (1971), p. 7.

and its SACs. The responsibilities of both were to formulate guidelines for training in the surgical/medical specialties, to approve posts and training programmes which were suitable for higher training and later to issue certificates of accreditation, indicating the successful completion of specialist training. These were tasks that the individual Colleges had undertaken independently in the recent past. But now, in the post-Todd world, there was a recognised need to create national inter-collegiate bodies to police standardised training more efficiently.[140] While the Joint Higher Training Committees were, as Norman MacKay recently commented, 'animals of the Colleges',[141] in that they were inaugurated by them and had a high number of collegiate representatives, they were another feature of the post-Todd system of postgraduate organisation that took things onto a new, higher, national plain. The Colleges still had immense influence in laying down higher training guidelines, but there was a sense in which some of the local bonds and involvements had been sacrificed to become part of the mainstream, national system.

Early in 1970 the MRCP Common Part 1 was held in Ceylon for the first time. This began a trend of overseas examining which was to become a characteristic feature of College activity in the next thirty years. Also, early in the year, word was received that the GMC had formally recognised the MRCP (UK) as a registrable qualification. The Common Part 2 examination was still being actively pursued.[142] In January 1970 the RCPL published a report based on a review of its own examinations. This was accepted as the basis for the new examination by both Glasgow and Edinburgh and joint meetings of all three Colleges went on throughout the year, with London continuing to set the agenda. Edinburgh accepted that it had now finally to give up the specialist option in its own examination, which was a major concession as this had attracted large numbers of candidates (with consequently high revenue). However, the RCPL decided, and the other Colleges accepted, that from June 1970, the MRCPs of all Colleges would include a

---

[140] The JCHMT was made up of representatives of the four Royal Colleges (RCPL, RCPEd, RCPSG, RCPI), the Professorial Heads of Departments of Medicine and Paediatrics, the Faculty of Community Medicine and various specialist associations, plus observers from the MRC and the government health departments for England/Wales, Scotland and the Republic of Ireland. The JCHST was made up of representatives from the four Royal Colleges (RCSEng, RCSEd, RCPSG, RCSI), the University Professors of Surgery and the specialist associations. See Robin Dowie, *Postgraduate Medical Education and Training: The System in England and Wales* (London, 1987), pp. 233–35; and *Joint Committee on Higher Medical Training: Training Handbook, 1990–91* (London, 1990). I am grateful to Dr Robert Hume for a copy of this publication.

[141] Interview by Andrew Hull with Professor Norman MacKay, 24 August 1998.

[142] *Council Minutes,* 13 January 1970, RCPSG, 1/1/3/18.

Paediatrics specialist option, though no others.[143] In May 1971 the three Colleges set up a joint Common Part 2 Planning Committee to agree the form, length, timing and scoring of the Common Part 2. The new examination dispensed with written essay answers entirely, as it was felt that the degree of marker variation meant that these were not an objective test of a candidate's knowledge. This section was replaced with a new tripartite section:

1. *Gray Cases*: The candidate is presented with the clinical history, findings on examination and certain investigations relating to an actual patient. He is then asked to answer certain specific questions: for example – What is the diagnosis? Give three likely diagnoses? What further investigations would you request? What treatment would you adopt? The candidate is given five such case problems, at least one of which will be a paediatric case, and is asked to answer any three of them.

2. *Interpretation of Data*: Laboratory reports, electrocardiograms, etc., are presented to the candidate and he is asked questions about their interpretation. Ten items are included in this section and all of these must be answered.

3. *Projected Material*: Slides of clinical photographs, radiographs, blood films, etc. (twenty in all) are projected on to a screen and the candidate has to answer specific questions about them.[144]

This part of the examination would be held simultaneously in all three examination centres, and the projected material viewed would be identical. This section would be taken by all candidates. The rest of the examination was to consist of one clinical and one oral examination. Candidates who had chosen to sit the examination in Paediatrics would have both their clinical and oral in that subject. Apart from the fact that the new-style written section taken by all candidates was to have at least one Paediatric case amongst the five gray cases to be chosen from, this was the only concession to specialisation. The new MRCP (UK) was an examination in General Medicine taken early in the career of the would be hospital specialist. It was an indication of potential and passport to higher specialist training. These new regulations came in in October 1972, and the Common MRCP (UK) Parts 1 and 2 were finally complete.

The colleges of Physicians had thus responded directly to the challenge of Todd by combining to offer a unitary, common examination. The colleges of Surgeons, who had been less comfortable with the idea of the FRCS as an indication of potential only, and not as an indication of a more complete

---

[143] See Craig, *History of the RCPEd*, pp. 774–76; Manderson et al., 'Evolution of an Examination', p. 104.

[144] R. L. Richards, 'The MRCP (UK)', *College Bulletin*, 2, no. 2 (July 1972), p. 12.

training, nevertheless were propelled along with the force of the current of thinking of the Todd Report, and the consensus between the Colleges on the early placing of examinations which arose from it. Increasingly, the trend was to consider the FRCS also as an early index of potential. The Surgeons had, however, their own way of expressing this in the late 1960s and early 1970s. As Andrew Watt Kay, then Visitor, wrote in the summer of 1971, a common examination was neither the only nor the best way of solving the problem of multiple diplomatosis:

> The four Royal Colleges of Surgeons in the British Isles firmly acknowledged the need to deter young men from sitting multiple Fellowship examinations but at the same time wished to retain flexibility, with an opportunity for agreed experiment in their respective examinations. The following statement appeared in the second report of the JCHST and will shortly receive much wider publicity: 'The four Royal Colleges wish it to be known that it is their unanimous view that it is normally unnecessary, and indeed undesirable, for trainees to obtain more than one of these Fellowships by examination. This mutual acceptance of the standard of each other's Diplomas is regarded by the four Colleges as a preferable alternative to the creation of a single Final Fellowship Examination for the British Isles.[145]

Kay believed that these examinations, taken after four years, indicated, like their medical equivalent, merely the potential to go on to further specialist training at Senior Registrar level. Looking ahead and speculating along the lines laid down by Todd, he suggested that this further training would then lead to the award, without further examination, of a Certificate of Specialist Training. However, the fact that the British Isle's higher surgical qualifications, unlike their European, American, Canadian, Australasian and South African counterparts, did not mark the end of specialist training, left a number unresolved problems. In 1971 Britain was gearing up to join the Common Market, and would then be subject to pan-European legislation on medical qualifications. This would affect both medical and surgical Examinations, and would be a problem that was finally solved only in the 1990s with the Calman Report and the Specialist Medical Order. Then another tier of higher examinations at the end of specialist training would be introduced by the Surgeons, who, in contrast to the 1970s, were then leading the way.

A feature of the Joint Higher Training Committees work that was to become increasingly important was also begun now. In 1970 the JCHST announced that it was to begin a system of accreditation of higher trainees. Those who were embarking on specialist training, after completing their Fellowship, would register with the JCHST, who would then grant them a

---

[145] Andrew W. Kay, 'The Present Position of the Surgical Fellowship', ibid., p. 8.

certificate of accreditation on the successful completion of that training. The scheme was voluntary, and hospital appointments committees were under no obligation to require certificates from those they wished to appoint as consultants, but the scheme was to be of increasing importance as in later decades attention turned from GPT to the formalisation of arrangements for the following period of specialist training.

At the same time, the early 1970s saw developments in the local administrative organisation of postgraduate education that left the Glasgow College with a slightly less hands-on role. Once again the Todd Report was the stimulus here. In 1970 the new national and local bodies to control postgraduate education were established: the Scottish Council for Postgraduate Medical Education (SCPGME) and the Western Regional Committee for Postgraduate Medical Education (WRCPGME). This was the end of the system, peculiar to Glasgow, where the Faculty/College and the University had together developed and controlled postgraduate education through their unique relationship. It was the beginning of a standardised, national administrative organisation along lines suggested by Todd. As R. B. Wright noted as President in 1970:

> While regional control and direction will to a large extent pass to the new Regional Committee on Postgraduate Education, our partnership with the University of Glasgow will continue and as a College we will still have particular contributions to make as a sponsor of lectures, courses, etc. The College will be represented on national and regional committees and will continue to act as guardian of high standards of professional behaviour and attainment.[146]

Elsewhere there were other indications of the new ways of working for the Glasgow College emerging after Todd. In 1972 the Faculty of Community Medicine (later the Faculty of Public Health Medicine) was established with representatives from all the Royal Colleges of Physicians. Instead of attempting to block the breakaway of specialists, after Todd, the Colleges sought to make a partial breakaway easier by facilitating the creation of a number of specialist Faculties. However, as in the above case, the new Faculties were in large part made up of College representatives and so a link was preserved with the traditional collegiate system. The changing role of the College, and perhaps particularly the somewhat diminished central role in the local administration of postgraduate education, made the Council look for other ways of raising the College's local profile as a professional organisation. One new initiative was the start of the *College Bulletin* in 1971. First edited by Dr J. Wilson Chambers, the magazine still survives as the chief organ of the

---

[146] 'President's Letter', *AGM Billet*, 2 November 1970, RCPSG, 1/1/5, p. 9.

College, and has recently been increased in size. As the then President, Edward McGirr, noted on the first page of the first issue:

> The Council of the College has become increasingly aware of a need to improve its communications with Fellows and Members, particularly those who do not reside in the West of Scotland and therefore cannot attend the monthly meetings regularly or participate in the professional, educational, and social activities of the College. The 'College Bulletin' has been introduced to overcome this deficiency and should in future help to keep Fellows and Members in touch with College views and news. The prestige and influence of an institution such as the Royal College of Physicians and Surgeons of Glasgow depend ultimately on the interest and involvement of its Fellows and Members in its activities.[147]

The College had spent the previous 120 years ensuring its fair representation in national developments in medical education. In the early 1970s this had been achieved to most dramatic effect, and the College was a full partner in a standardised national system of postgraduate education. It was therefore, perhaps fitting that it now turned back to reassure its Members and Fellows that it had not forgotten that they were, fundamentally, the College itself.

[147] 'President's Column', *College Bulletin*, 1, no. 1 (1971), p. 3.

6

Towards the Future, 1972–1999

Medicine used to be simple, ineffective and relatively safe. It is now complex, effective and potentially dangerous. The mystical authority of the doctor used to be essential for practice. Now we need to be open and work in partnership with our colleagues in health care and with our patients.[1]

We are truly, not merely a College, but more really an Academy, representing all facets of Dental, Medical and Surgical practice.[2]

The College is no longer a parochial institution, interested only in the West of Scotland. One has only to attend a Ceremony of Admission and walk among the new Fellows and Members to see the spread of countries from which they are drawn.[3]

Since 1972 the College has continued to function as one of the major UK postgraduate examining bodies in medicine and surgery. It is a professional body with a fundamental commitment to ensuring that hospital doctors are trained and examined to the highest possible standards in the most modern conception of what constitutes medical knowledge.

In this most recent period, postgraduate education has continued to develop along the lines laid down in the Todd template. There have been important developments in the nature and scope of examinations, and in the local and national machinery for administering and overseeing postgraduate training and education. These developments have meant further evolution of the role of the College. Readers of this book will realise by now that this is a continuous process because medical knowledge is itself constantly evolving, and because of the interplay of this knowledge with social factors such as the priorities of legislators, the attitude of the public towards medical practice and provision

---

[1] Sir Cyril Chantler, Dean of the University Medical and Dental School of Guy's and St Thomas's Hospitals (London's largest medical school), speaking in America in October 1998. Cited from 'The End of Secrecy: New Medical Ethic Needed', editorial in the *Guardian*, 23 November 1998, p. 17.

[2] 'President's Letter' (from Mr Ian A. MacGregor), *AGM Billet*, 4 November 1985, RCPSG, 1/1/5. All billets are in this location.

[3] Ibid.

and the profession's own perception of its identity. There have also been important developments in the internal administrative structure, and external relations of the College in this latest period. Furthermore, an extra dimension has been added to Collegiate activities with the evolution of a programme of involvement in both the examination of overseas candidates and the provision of high-quality training abroad.

In advance of the reorganisation of the National Health Service in Scotland from 1 April 1974 (which, among other reforms, split RHBs into more numerous Health Boards),[4] the organisation of postgraduate medical education was altered in 1970. The idea was to create a new regional controlling body along the lines laid down in the Todd Report. After 1974 six new Health Boards assumed responsibility for the area covered by the old WRHB. In postgraduate education, this meant that the University of Glasgow had to associate with six area postgraduate committees, instead of one regional postgraduate committee, the old GPGMB, formed in 1960. Coordination was aimed at with the establishment in 1970 of the Western Regional Committee for Postgraduate Medical Education (WRCPGME), a non-statutory body with a remit for the whole of the old region. The WRCPGME differed from the old GPGMB in that it included representatives from Health Service authorities and other Royal Colleges and Faculties.[5] Under the new administrative arrangements, however, there was no statutory obligation for the six individual Health Boards to entrust the organisation of postgraduate education to any regional committee, though this was, of course, the wish of the professional bodies. The new Health Boards might well wish to administer the postgraduate training of doctors in their area themselves.[6]

In 1974 the regional committee was renamed the West of Scotland Committee for Postgraduate Medical Education (WSCPGMEC) to distance it from the old RHB organisation, and to emphasise its remit for postgraduate education in the whole of the west of Scotland. The Health Boards now followed the opinion of most of the doctors in the west of Scotland and allowed this new committee overall administrative control of postgraduate medical education.[7]

---

[4] Under the *National Health Service (Scotland) Act,* 1972.

[5] Details from SC(73)11, 'Scottish Council for Postgraduate Medical Education: Organisation of Postgraduate Medical Education in an Integrated Health Service. Report of a Working Party', 10 May 1973, Shaw Papers. Shaw was a member of this working party.

[6] See 'Minutes of the Thirteenth Meeting of the Western Regional Committee for Postgraduate Medical Education Held in the University of Glasgow on 11 October 1973', point 29, 'The Future of the Regional Committee', Shaw Papers.

[7] Ibid. See also Kenneth Calman, 'Continuing the Conversation: A Look into the Future', in Dow and Calman (eds), *RMCSG: A History,* pp. 115–22.

The new organisation meant a revision of the role of the College. Whereas, since the University's submission to the NPHT in 1962 for funds for post-graduate organisation in the wake of the Christ Church initiative, the College had been the acknowledged regional postgraduate centre, the policy was now to establish a variety of local postgraduate centres, to provide a more local focus and access to resources for postgraduates in the larger geographical area. For example postgraduate centres were established on the Stobhill and Southern General Hospital sites. Moreover, the University of Glasgow had become the natural home for the offices of the (pre-1974) WRCPGME, for practical reasons of space, and also because the University Postgraduate Dean was the pivotal figure in the regional organisation envisaged by Todd.[8] In 1972, however, a working party of the WRCPGME thought long and hard about a central role for the College, eventually rejecting the idea because the College was not concerned with the postgraduate education of general prac-titioners, which any regional committee would have to be. The working party,

> did consider the advantages and disadvantages of developing the Royal College of Physicians and Surgeons as a sole centre. There is no doubt that it has already become an acknowledged educational focus for local and national events and pro-vides a splendid central medical library [much better than the University's]. It was felt, however, that unless some adjacent or nearby premises could be found which would be widely used by general practitioners – and this was considered unlikely – the present Royal College would remain essentially a College of hospital spe-cialists, therefore unlikely to achieve the full purpose of the proposed centres in regularly bringing together doctors of all branches of the National Health Service.

Rather, it was decided that the College should retain an important role as an intellectual centre in a new mix of postgraduate provision:

> *We feel that a centre of medical education of the size of Glasgow requires the facilities both of the Royal College and the development of local centres.* The Royal College should be made fully capable of providing a central meeting place for regional and national events and in developing central library and audio-visual facilities, having a close and agreed working relationship with university and hospital library services.[9]

[8] The Postgraduate Dean, Charles Fleming, while a strong supporter of the College's new role, was quite clear about the centrality of the University in the new system of postgraduate organisation. As he wrote to Shaw, 'I am, of course, anxious that no one should be in any doubt about the Glasgow view that the postgraduate office should be in the University ...' See Fleming to Shaw, 10 March 1972, Shaw Papers.

[9] EC(72)4/RC(72)4, 'Western Regional Committee for Postgraduate Medical Education: A Report on Postgraduate Medical Centres in Glasgow and the West of Scotland 1972', March 1972, Shaw Papers. Shaw was also a member of the working party that produced this report. Original emphasis.

Underlining its new role as the intellectual centre of postgraduate education in the west Of Scotland, in 1977 the College established an Education Committee to organise all educational events.

So the College had a slightly reduced, less hands-on role now in postgraduate education, but remained, as it still does remain, the intellectual centre. It retained its close involvement with the formulation of guidelines for specialist training, the approval of posts and the accreditation of those who had completed higher training, through the JCHMT, JCHST and the Joint Committee for Higher Training in Dentistry (JCHTD). It also, of course, continued to offer its own examinations, most notably the Membership and Fellowship.

By about 1971 the Glasgow College, in line with the other UK Royal Colleges, had largely accepted that both the Membership (MRCP UK) and the Fellowship (FRCS) signalled the end of GPT, 'both as a measure of clinical competence and as a test of future potential'.[10] Though the Surgeons are still to take the final step in this direction, the achievement of the common MRCP (UK) in 1972 marked a shift of concern from these higher examinations, taken early in the hospital career as an indication of potential, to a focus on how the end of Specialist Training (ST) should be assessed. Initially the Joint Higher Training Committees marked the successful completion of ST by the issue of a certificate of accreditation. But this was a voluntary scheme and hospital appointments committees were not bound to employ only accredited candidates as consultants. The development of mechanisms for the assessment of Specialist status was one of the major themes of the period from 1972. There were also a variety of changes in the higher examinations marking the end of GPT.

The College had supported the idea of a Specialist Register held by the GMC for many years. This position was reemphasised in the College Working Party's submission to the Merrison Committee during 1974. The College wished to see the GMC exercise overall control over postgraduate vocational and specialist qualifications. The Joint Higher Training Committees would define higher training programmes and advise the GMC on who was qualified to go on the Specialist Register. The Merrison Report of 1975 led to the Medical Act of 1978.[11] This made two additions to the GMC's statutory responsibilities: a general function of ensuring high standards of medical education; and a responsibility for overseeing *all* stages of medical education. This meant that now the GMC stood in the same relationship to the Royal

[10] Edward McGirr, 'President's Letter', *AGM Billet*, 1 November 1971.
[11] *Report of the Committee of Inquiry into the Regulation of the Medical Profession*, Cmnd 6018 (London, 1975).

Colleges and Faculties in regard to postgraduate and continuing education, as it did in relation to the medical schools in regard to undergraduate education. The GMC already had an Education Committee, but now it was established as an independent statutory committee with specified functions and legal powers. It was now this committee, and not the Council as a whole, which was required to attend to educational standards, and the Council's powers of visitation, inspection and access to information were transferred to it.[12] The Act did not require the GMC to hold a Specialist Register, but indications of specialist competence (specialist or additional qualifications) could be shown in the existing general Medical Register. Thus there was still no indication of whether persons styling themselves specialists in a certain field had actually had any training or experience in it.

The issue was, however, becoming increasingly pressing. Britain had entered the Common Market under the Conservatives in 1973, and membership was confirmed by the referendum called by the Labour Government in 1975. Also in 1975 a European Directive made certification of specialist qualifications and the holding of national Specialist Registers mandatory.[13] However, from 1978 to about 1983 the Education Committee (after discussions with the Royal Colleges and Faculties, and the University Postgraduate Deans and others) decided to concentrate its focus on GPT (or what it came to call, reflecting the dominance of the specialising impulse, 'Basic Specialist Training').[14] Margaret Stacey, who served on the GMC as a lay member, with experience of medical administration, from 1976–84, has offered an insider's perspective on the reasons for this focus. Under the post-1978 arrangements, the Education Committee's perspective was informed by two main influences:

> First, there was a wish to establish a pattern of working relations with the royal colleges and others before recommending anything very drastic – the requirement imposed by professional consensus. Second, the Council was nervous throughout the 1980s of proposing legislation; the political atmosphere was not favourable to professionals in general and to medicine in particular, suggesting that there might be above-average hazards in requesting legislation – better to hope the dogs would snooze if not disturbed; clearly they were not very sound asleep.[15]

The problem of the definition of a specialist nevertheless had, as noted many times already, important implications for the whole of the postgraduate

[12] Margaret Stacey, *Regulating British Medicine: the General Medical Council* (Chichester, 1992), p. 118.

[13] See EEC Directive 75/362/EEC.

[14] See Dowie, *Postgraduate Medical Education*, pp. 264–65.

[15] Stacey, *Regulating British Medicine*, p. 121.

structure. Would the Todd template – with its progression from GPT to ST – hold? This question was of particular importance in Surgery, which, unlike Medicine, had not achieved a common examination structure for the end of GPT. In 1977 the Edinburgh Surgeons proposed to establish a specialist Fellowship, thus threatening to unravel the Todd consensus on the nature, scope and timing of postgraduate examinations. The London College strongly opposed the move. In Glasgow, there was a mixed response. Tom Gibson wrote as President that:

> while we have felt for some time that some method of recognition of specialist qualifications was desirable ... of course, our position is more complex than that of the other Colleges because of our joint status. Throughout the English-speaking world, however, there is no doubt that Britain is the odd man out in not having some form of examination to test qualifications in the major medical and surgical specialties.[16]

The situation in Surgery fragmented further in 1980 when reciprocity between the Primary Examinations for Surgical Fellowships (negotiated in 1951) was lost.[17] The Edinburgh Surgeons were pursuing their independent idea of a specialist Fellowship. The Glasgow College was running MCQs but no essay questions, a format that the RCSEng was not comfortable with, and it was the English College that withdrew from the reciprocity agreement.

A new cooperative strategy, however, soon began to emerge between the Surgical Colleges. Cooperation, rather than competition, between the Colleges and also with the specialist associations was, as noted above, increasingly a feature of British medical education, especially after the Todd Report. This tendency was reconfirmed with the achievement of the MRCP (UK). By 1984 the Colleges were discussing the concept of an intercollegiate specialty examination in Surgery. There was little hope of quickly reformulating the existing examination structures, yet such an examination, administered jointly by all the Surgical Colleges in the British Isles, was up and running in Plastic Surgery by 1986,[18] and by 1988 in Neurology, with examinations in Orthopaedic and General Surgery on the horizon.[19] By 1994 the Intercollegiate Specialty Fellowship (ICSF) had been established in the areas of General Surgery, Paediatric Surgery, Plastic Surgery, Orthopaedic Surgery, Otolaryngology, Oral and Maxillo-Facial Surgery, Surgical Neurology, Cardiothoracic Surgery and Urology.[20] By 1991 all surgeons had to pass

[16] Tom Gibson, 'President's Letter', AGM Billet, 7 November 1977.
[17] See ibid., 3 November 1980.
[18] I. A. McGregor, 'President's Letter', ibid., 4 November 1985.
[19] Alistair D. Beattie, 'Honorary Secretary's Report', ibid., 6 November 1988.
[20] 'Notices' item 10, 'Intercollegiate Specialty Boards', ibid., 7 November 1994.

the ICSF before being accredited, and the examination was registered by the GMC.[21]

By 1992–93 the problem of harmonisation of British and European specialist training times and examinations denoting specialist status had become acute.[22] The Calman Committee was established to suggest possible solutions to this problem, and to formalise specialist medical training more generally. The Calman Report[23] recommended, 'a curriculum for each specialty, structured training programmes, progression through training based on formal annual assessments of competence, and much shorter training in most specialties'.[24] The new system was to be provided by the Royal Colleges and the Postgraduate Deans by the end of 1995, and a new grade of Specialist Registrar replaced those of Registrar and Senior Registrar. On 12 January 1996 the European Specialist Medical Qualifications Order passed into British law.[25] This order established the Specialist Training Authority as the 'competent authority' to issue, after the completion of the appropriate training programme, Certificates of Completion of Specialist Training (CCSTs). These certificates – which were the direct descendant (and a formalisation) of the old system of accreditation – were to be registered with the GMC on a new Specialist Register. After 1 January 1997 no doctor could take up a consultant post in the NHS unless his or her name appeared on the Specialist Register. In Surgery, the appropriate ICSF was made an integral part of the getting of the CCST in any particular specialty. In Medicine there is as yet no comparable exit examination to Surgery's ICSF, but rather a system of continuous assessment operates leading to the granting of the CCST by the STA. This assessment is overseen by the Colleges via the JCHMT. However, this issue of an exit examination in Medicine is something which is currently under discussion by the Colleges of Physicians, particularly after recent events, and the accompanying public and professional sensitivities about the legal definition of a specialist. The current thinking among the Colleges of

[21] Alistair D. Beattie, 'Honorary Secretary's Report', ibid., 6 November 1989.

[22] This was due to a particular case coming before the European Court which exposed the problem of British recognition of European specialist qualifications, and the European problem of accepting British specialist training times. Britain had generally longer training times but required less specialist experience.

[23] Department of Health, *Hospital Doctors: Training for the Future* (London, 1993). The chairman was Dr Kenneth Calman.

[24] John Biggs, 'New Arrangements for Specialist Training in Britain', *BMJ*, 311, 11 November 1995, pp. 1242–43. Quotation from p. 1242.

[25] *Statutory Instruments*, 1995, no. 3208, 'Medical Profession: The European Specialist Medical Qualifications Order 1995'. On 1 January 1997 Article 11 and Schedule 5, dealing with mandatory possession of CCSTs for NHS hospital consultants, came into effect.

Physicians is to make the continuous assessment of candidates more rigorous rather than to introduce a formal exit examination.

The postgraduate training and examination of specialists had finally been formalised along the lines suggested by the Todd Report by 1997. This, of course, is far from the end of the story. During 1998 in particular, a number of problems arose in medical (and especially) surgical practice with regard to the quality of medical and surgical care delivered. These were seized on by the media (which was able to link them in nicely to the fiftieth anniversary of the NHS). This created a heightened public (and professional) sensitivity towards the qualifications of hospital doctors. Most obviously there were the serious issues of surgeons' accountability inherent in the deaths of (possibly *at least*) twenty-nine babies and toddlers between 1995–98 as a result of heart surgery at the Bristol Royal Infirmary, but there was also a continuing rash of (possibly heterogeneous) problems with medical care coming to light on a seemingly weekly basis.[26] Furthermore, Dr Richard Kaul, an anaesthetist, who had completed his consultant training in the United States, returned to Britain (coincidentally to the Bristol Royal Infirmary, Chronic Pain Relief Unit) to find that he was required to undertake a further year's training in order to qualify for the equivalent consultant status that he had enjoyed in America. He promptly issued a legal challenge to the STA, questioning the basis of their decision in his case, and this case is still *sub judice*.[27] The Senate of Surgery (a joint body representing all the Surgical Colleges and the surgical specialist associations, founded in 1994) responded specifically to the Bristol case, after the GMC's enquiry had reported on 18 June 1998 finding three Bristol doctors guilty of serious professional misconduct.[28] The Senate's recommendations were as follows:

1. Methods of revalidation should be established for consultants.

2. Efficient systems to gather activity and outcome data should be developed with the help of Colleges and associations on a national basis, using the proposed developments in information technology. This is a complex, costly and lengthy task, requiring significant investment.

3. The Department of Health with the assistance of the Colleges and associations should arrange a 'rapid response team' to provide expert external assessment within two or three days when there is any suspicion that patients' safety may be at risk. Every effort should be made to avoid suspension, pending a full inquiry.

[26] See, for example, Jeremy Laurance, 'Doctor, Heal Thyself', *Independent on Sunday*, 4 October 1998, p. 23.

[27] Sarah Boseley, 'Doctor Plans First Challenge to Surgeons' "Closed Shop"', *Guardian*, 29 June 1998 (front page story).

[28] GMC Professional Conduct Committee, *Announcement of Determinations by the President of the GMC* (London, 18 June 1998).

4. The independent practitioner status as defined in the consultant contract should be reviewed. The aim should be to encourage consultant team working and to facilitate clinical governance. However, this must be done without sacrificing the advantages which arise from personal responsibility for individual patients or the safeguards and the opportunities for innovation which are conferred by clinical freedom.

5. The annual assessment of surgical trainees needs to be more rigorous, especially in early years.

6. Further research should be done to clarify the causes of poor performance and professional conflict. Methods need to be developed to foster an open culture, cooperative working, personal insight and skills in communication.[29]

The key recommendation here, so far as the changing future role of the Colleges is concerned, is the endorsement of revalidation (or reaccreditation). This would probably be done every five years, either by inspection, clinical audit, assessment or examination, and will involve the Colleges in a huge new bureaucratic and conceptual effort. However, the Labour Government has already issued a White Paper on Health suggesting the establishment of a National Standards Board (in England a Commission for Health Improvement), a new agency, resembling a kind of 'OffDoc', to inspect the competence of units and publicise those with unacceptable standards, so there will undoubtedly be further developments in this area. It appears that, once again, the Colleges will have to defend and amend the British system of self-regulation of the medical profession. As an editorial in the *Guardian* recently commented, perhaps reflecting the contemporary climate of public opinion: 'Self-regulation has an unhappy history, but now that it is supported by independent monitoring, it should be given one last chance'.[30] The point is, of course, that, as all doctors know, medical education is a constantly evolving concept. It is influenced not only by advances in medical knowledge but also by wider cultural and social changes. What can or does a society expect from its doctors?

There have also been a number of important developments in the College's Membership and Fellowship (and other examinations taken before ST) in this period. In 1986 the numbers sitting the Final FRCS Examination topped one thousand for the first time, and in 1987 and 1991 new College diplomas were introduced in, respectively, Ophthalmology and Sports Medicine (the latter a joint diploma with the other Scottish Royal Colleges).[31]

---

[29] The Senate of Surgery of Great Britain and Ireland, *Response to the General Medical Council Determination on the Bristol Case* (London, 1998), p. 9.

[30] 'The End of Secrecy', *Guardian*, 23 November 1998, p. 17.

[31] See *AGM Billet*, 3 November 1986, 2 November 1987, 4 November 1991.

There has been a renewed effort among the Surgical Colleges to work towards a common surgical examination to compliment the MRCP (UK). This had always been a particular concern of Sir Robert Wright as President from 1968–70. He felt that the title of FRCS was inappropriate for an examination taken early in the career of the hospital doctor, and that all the UK Surgical Colleges should institute a common MRCS as the gateway to specialist training.[32] In 1996 the Glasgow College began its own MRCS taken after two years surgical training as the mark of the end of GPT and an essential step on the route to higher surgical training. The MRCS differs from the old FRCS in that progress during training is much more closely monitored and candidates are required to pass a basic surgical skills course before sitting the examination. It has not yet proved possible to negotiate an intercollegiate (common) examination. Each College holds its own examination (Glasgow and London offer an MRCS, Edinburgh and Dublin an AFRCS), but reciprocity has been agreed. The first two diets of the MCQ component, in applied basic sciences and clinical surgery, took place in May and September 1997, and the first sitting of the orals and clinicals followed in the spring of 1998. There is a transitional period when the old-style FRCS examination is being retained for those candidates already in training, but this will end in 2000. The College anticipates that the numbers sitting the MRCS will be lower than those for the FRCS, because of the increased entry requirements. It is likely that very soon a common MRCS will be achieved, and certainly this is what ongoing negotiations between the Surgical Colleges are aiming towards. As the President, Colin MacKay, recently commented:

> I have always been anxious that we would have an MRCS (UK and Ireland). I accept that this would probably be to our financial disadvantage but I do feel that the MRCP (UK) has been a success, but not without its problems. I would like to try and develop the good things about the MRCP and avoid the bad things when we design our MRCS and I would love to see the Irish College on Board so it would be an MRCS (UK and Ireland). This would be an example to Europe and to the world, and hopefully, if we could see their training was to the same level as ours, we could have people from other countries sitting it too. I am glad to say that I do see some glimmer of light at the end of the tunnel. We've been talking about this for many years now but we are coming closer together and I would hope that in the foreseeable future we might end up with that.[33]

In dentistry, the T. C White Bequest of £220,000 in 1982 enabled the

[32] See Gavin Shaw, 'Sir Robert Wright, DSO, OBE', *College Bulletin*, 11, no. 2 (January 1982), pp. 8–9.
[33] Interview by Andrew Hull with Mr Colin MacKay, PRCPSG, 11 September 1998, RCPSG, 1/13/9.

establishment of schemes for postgraduate training. The GDC and the Joint Committee for Specialist Training in Dentistry (JCSTD: an updating of the JCHTD) have been defining specialist training programmes in dentistry, and during 1997 an accord was reached as a basis for cooperative action in this area between the GDC, the Dental Faculties of the Colleges, the Specialist Societies, the Universities, the Postgraduate Dental Deans and Directors and the JCSTD. The Dental Surgeons have also followed the system of postgraduate examinations adopted by the Surgeons in establishing a new MFDS in 1998 to precede the new intercollegiate specialty examinations in Dentistry. As in Surgery, the old style Final FDS will be phased out in 2000 to allow a transition period for those already in the examination process.

The achievement of a position of equal respect and status for the College as a professional medical examining body has been both accompanied and advanced by a keen awareness that the internal administrative structure must be constantly kept under review and periodically modernised. The changing role of the College has also meant that the external relations of the College with newly emerging specialist organisations, national medical organisations and other professional groups has also changed.

The clearest indicator of the College's changing professional priorities and its perception of its professional role is the composition of the executive heart of the College: the College Council. The Council was established as the College's executive in October 1850 as part of the contemporary general modernisation strategy. The modern history of the College begins with the Faculty Act of 1850 which aligned the Faculty with the progressive movements towards medical reform, thereby beginning the process of ensuring the Faculty a place at the top table of medical licensing bodies. As part of this process the internal administrative structure was tightened and made more efficient in order to make the best of the Faculty's new position, and to reflect its ambitions. Three resident Fellows were now to be elected annually by the full Faculty at the AGM (then in October). These, along with the President, the Visitor and the Honorary Treasurer would form the 'Council', which would take care of 'all ordinary business' and would have the power to add to its number for the consideration of special questions.[34]

The expanding role of the College in postgraduate education led to the creation of a new post in 1953, that of Honorary Secretary, who was also a member of Council. In 1990 representatives of the Faculty of Public Health Medicine and of the Faculty of Pharmaceutical Medicine were added to the Council.[35] In 1997 a representative from the new Faculty of Accident and

[34] *Faculty Minutes*, 7 October 1850, 'Regulations', 'Council', RCPSG, 1/1/1/8.
[35] *AGM Billet*, 5 November 1990.

Emergency Medicine was also added to the Council. These representatives served a dual purpose as they were reciprocally the College's representative on the Council of the respective Faculty, and representative of their Faculty on the Council of the College. This is also an important indication of the new *modus vivendi* of the College. It is an umbrella organisation but with strong, permanent two-way links to the new Faculties. The pattern of the new relationships is well demonstrated by I. A. McGregor's 1985 presidential comment on the relationship of the College with its Dentists:

> Our Dental component continues to flourish led by the Dental Council, managing its own affairs within the overall framework of the College, lending strength by its presence, contributing to our realisation that we are truly, not merely a College but more nearly an Academy, representing all facets of Dental, Medical and Surgical practice.[36]

The Dental Fellows attained full voting rights within the College in 1981, and in the Spring of 1982 the Membership in General Dental Practice was established.[37]

As President between 1991–93, internal administrative reform was a particular priority for Robert Hume. For him the reform of the Council was a way of keeping the 'Academy' intact, of retaining specialists within the orbit of the College by giving them influence over policy. He wished to see a parallel 'Specialty Advisory Council' created within the College to compliment the existing executive Council, and to bind specialists further into the College structure. As he recently commented:

> I wanted to see it [the Council] expand by specialist representatives so that we could feed out into the community and write reports on any specialty, or that the College would be the voice that, for example in medical practice, said what was the best practice for this or that, and all this would emanate from *College*, and not from the Scottish Office, or other groups outside, because they would all be *here*, and we would be the *hub* of activity.[38]

This change would have reinforced the other important change that Hume oversaw, the appointment of RCPSG representatives or tutors to all the local hospitals,[39] a communications strategy that, as noted earlier, the RCSEng had adopted as early as 1961. This established a two-way link between the Glasgow College and the local hospital world, which was crucial for the

[36] I. A. McGregor, 'President's Letter', *AGM Billet*, 4 November 1985.

[37] *AGM Billet*, 2 November 1981.

[38] Interview with Dr Robert Hume at RCPSG, 10 September 1998 by Andrew Hull. Original emphasis.

[39] Ibid.

coordination of training and for feedback from specialists and others. Again, it was a way of keeping the 'Academy' together, and at the 'hub' of things.

This specialist council did not come to pass, but Hume did successfully oversee two further reforms of great importance. The presidential term was extended to three years, the Visitorship reduced to one year and two new posts of Vice-President (one in Medicine and one in Surgery) were created to ease the burden of the President's ever-expanding workload. These posts would be held for two years and were to be filled by a Fellow *qua* Physician and a Fellow *qua* Surgeon, respectively. For a College of Physicians *and* Surgeons, this move made a great deal of sense.[40] As Hume recently commented:

> We were not pulling our weight or making an impact I think in the same way as the other Presidents. There was a perception in London certainly by the President of the Surgical College who said he thought we were just two half Colleges therefore our representation on Committees should be half the Edinburgh College and a third of theirs because they thought they should have more representation, sort of pro rata. I tried to point out to him that we were two Colleges under the one roof. The College President here has to run round the surgical circuit of meetings and the medical circuit of meetings and he's also to be at the combined meetings of surgical and medical Colleges, so he works twice as hard as the other Presidents who are looking after single Colleges. Therefore, I thought it would be much better if instead of a two-year Visitor we had a one-year Visitor and a three-year President and to increase the administrative staff by having two Vice-Presidents, one surgical and one medical, who could assist the incoming President and do some of the local chores and release him from the burden of duties. The President here had to do everything, open all the meetings, be everywhere at once. Open conferences he wasn't taking part in. I introduced it in 1992 against a lot of opposition because the senior people and past Presidents thought this was unnecessary but the volume of work had increased between Europe, junior staff, new training structures.[41]

The other reforms at this time were the establishment of the post of Assistant Honorary Secretary to aid the Honorary Secretary with his ever-increasing workload, the creation of the Junior Advisory Committee, and the appointment of a Senior Registrar to each of the College committees. As Hume commented at the time, both of these latter moves, 'added freshness and strengthened our links with the "College future"'.[42]

Most recently the Council has been expanded in a different way to both reflect and underline a *no less important* aspect of the College's mission. In 1997, under the presidency of Norman MacKay, seven Regional Councillors

[40] Robert Hume, 'President's Letter', *AGM Billet*, 2 November 1992.
[41] Interview with Dr Robert Hume at RCPSG, 10 September 1998, by Andrew Hull.
[42] Robert Hume, 'President's Letter', *AGM Billet*, 4 November 1991.

were added to the Council, both emphasising and helping to ensure that the College is a body representing the *whole* of Scotland and also the whole of Great Britain. The new Councillors were to be three Physicians or Surgeons from across Scotland (Grampian, Highland and Tayside combined regions; from the Borders, Fife and Lothian; and from west and central Scotland excluding Glasgow), together with a Physician and a Surgeon each from Northern Ireland, and from England and Wales. These new Regional Councillors would be elected by the Whole UK Membership and Fellowship. An earlier example of emphasising the national constituency of the College began in 1972, when the College began holding joint symposia in other Scottish cities. The first was on Gastroenterology in Aberdeen, and this has been followed by other symposia in, for example, Inverness, Dumfries, Oban, and, indeed, even as far afield as Madras. Also added to the Council, *ex officio*, in 1997 were the convenors of the Medical and Surgical Advisory Committees. Furthermore, of the ten existing 'ordinary' Councillors, five would henceforth be Physicians and five Surgeons (and this included one young Medical and one young Surgical Councillor). As Stefan Slater, then Honorary Secretary, commented, 'This is a constitutional safeguard against any gross imbalance in the numbers of Physicians and Surgeons on Council'.[43]

In order to make allowances for the new Councillors, some of whom have to travel long distances, and also for the increasing clinical commitments of Councillors, the timing and frequency of Council and College Meetings were changed in 1996. Council Meetings were made bimonthly instead of monthly, and College Meetings were made quarterly (in March, June, September and, as the new AGM, December).[44]

There have also been changes to the line up of national medical bodies on which the Glasgow College pools its views and exchanges opinions with other professional medical organisations. For example, in 1973 the Joint Conference of Colleges and Faculties in Scotland was formed. The first meeting was in April 1973, and the first Honorary Secretary of the new body was the College's own Thomas J. Thomson. The composition of the new body was one representative from each of the RCPSG, RCPEd, RCSEd, RCOG, Royal College of Pathologists, Royal College of Psychiatrists, Faculty of Anaesthetists, Faculty of Community Medicine (later, of Public Health), Faculty of Radiologists, and one observer from the Scottish Council for Postgraduate Education. Each representative was to act in turn as Chairman of the Conference for six months to reinforce a sense of unity and involvement. With a relatively small membership of only eleven there was

---

[43] Stefan Slater, 'Honorary Secretary's Report', *RCPSG Annual Report*, 1996/97, p. 8.
[44] Ibid.

scope for talk, exchange of views and agreement and thus, as Thomas Thomson himself commented in 1974, 'this body is able to express a Scottish viewpoint on standards and practice and training, not only in this country, but also in the European Economic Community'.[45] Its purpose was to ensure uniform methods of training and standards across all Scottish medical corporations and regions.[46]

On other fronts, this period saw the purchase of 234 St Vincent Street in 1975, and of 232 St Vincent Street in 1989, so that the College had by then expanded to its present considerable size. Finances had been steadily improving from the mid 1970s, and by 1978 a healthy surplus of £2869 was recorded. On 1 July 1983 Her Majesty the Queen visited the College, and in February of that year HRH the Princess of Wales agreed to become the Patron of the College. Diana visited the College to be made an Honorary Fellow in May 1984, again on 20 April 1989 and, finally, on the occasion of the opening of the Princess of Wales Conference Suite on 8 June 1993. In 1994 the College established a Court of Patrons composed of eminent local persons willing to assist the College in maintaining a high profile within the local community, and especially the professional community. Initial members included Lords Gould and Macfarlane, former Lord Provost of Glasgow, Dr Susan Baird, and Dr Jean Morris.[47]

One other event during this period which should be recorded was the death of Mr John Robb, the Registrar, while on College business in Singapore in 1990. The current President, I. A. McGregor described in 1985 the special importance of Mr Robb to the College: 'The custodian of continuity, the high priest of precedent, is Mr Robb. He carries with him the "folk memory" of past Council decisions. If Cardinal Richelieu had Father Joseph as his *eminence grise*, the College has Mr John Robb'.[48] No more needs to be said of his importance to the College, than that the Honorary Fellowship was conferred on him on 19 October 1988.

The third area in which there have been important, and characteristic developments in this most recent period, is in the beginning and expansion of overseas examination. The College's involvement in examining overseas had begun in 1970 when the Common MRCP Part 1 was held in Ceylon, but it was not until the early 1980s that this kind of activity really took off.[49] In

---

[45] Thomas Thomson, 'Conference of Royal Colleges and Faculties in Scotland', *College Bulletin*, 4 (September 1974), pp. 5–7.

[46] Sir Andrew Watt Kay, 'President's Letter', *AGM Billet*, 4 November 1974.

[47] See *AGM Billets* for 1975, 1978, 1981, 1983, 1989 and 1994 for the details cited in this section.

[48] I. A. McGregor, 'President's Letter', *AGM Billet*, 4 November 1985.

[49] All details in this section are taken from the President's or the Honorary Secretary's Reports in the relevant *AGM Billet* or (from 1995/96) the *Annual Report*.

April 1982 the Primary Surgical Fellowship was held in Tripoli. The College had been invited to hold the examination in Abu Dhabi in the same year, but this was abandoned due to a lack of anatomical specimens. During 1983 the Primary FRCS was held in both Tripoli and Abu Dhabi, and plans were afoot for holding the examination (along with the Primary FDS) in Kuwait in May 1984. During 1983 the College introduced an innovative sponsorship scheme for overseas graduates training for and taking examinations in the UK. The aim was to ensure that the training of overseas graduates was as organised, and up to the same standards, as that of home graduates. Overseas graduates were to be supervised in approved posts while undertaking study for Membership or Fellowship examinations. The trainees were then to be allocated to approved higher training posts currently surplus to NHS man-power requirements. At the end of such specialist training, the trainees would take an additional examination offering a diploma acknowledging specialist status equivalent to UK accreditation. The scheme was promoted by ex-President R. B. Wright when President of the GMC, with input from the newly formed College Overseas Committee and Norman MacKay, George D. Forwell (Chief Area Medical Officer), and Professor G. C. Timbury, then the Postgraduate Dean. All trainees had to have either the Primary FRCS or Part 1 of the MRCP (UK) to be eligible to join the scheme.

This scheme demonstrates College's concerns with training standards for overseas graduates. Likewise, overseas examination was a two-way process. It was a lucrative market for the College's prestigious qualifications, but the College was also adamant that there should be no drop in standards. Perhaps the clearest example of this concern was the Oman project. Here the College became intimately involved in establishing a modern infrastructure of medi-cal provision, as the basis for effective training to take place. The roots of College involvement in Oman can be traced back to Mr Campbell Semple, a Fellow of College, who built up contacts during a period of National Service in Oman. An exchange scheme in Orthopaedics developed some years later from these initial contacts and Omani doctors received higher surgical train-ing in Glasgow. At the same time Senior Registrars in Orthopaedics were seconded from Glasgow and Edinburgh to Oman for periods of six months.[50] In September 1983 the Omani Minister of Health, Dr Mubarak Al Khaduri, visited the College, 'to discuss the methods by which we might work together to further the development of medical services in Oman'.[51] The Minister identified various areas where Glasgow could be of particular assistance,

[50] Details in this section are from T. J. Thomson, CBE (PRCPSG 1982–84), 'The Oman Project', College Bulletin, 15, no. 3 (May 1986), pp. 13–16. I am grateful to the author for alerting me to this article.
[51] Ibid.

including Anaesthetics, Medical Laboratories, Nursing, Obstetrics and Gynaecology, and Surgery. In the next two years surveys of the existing facilities in these areas (plus Dentistry) were carried out. In 1970 the only medical provision in the sultanate was one hospital with twelve beds, nine health centres and ten dispensaries. By 1986 the Health Ministry oversaw fifteen hospitals and twenty health centres with 2133 beds, along with seventy-two dispensaries. In 1987 a fully equipped modern hospital facility, the Royal Hospital, was due to open.

The Glasgow College was closely involved in these developments, staff being seconded to Oman from Glasgow hospitals with the cooperation of the Greater Glasgow Health Board. In November 1984 a Director of Laboratory Services for Oman was appointed and he attracted and supervised the secondment of many medical laboratory scientific officers in biochemistry, blood transfusion, haematology, histopathology and microbiology, while medical and paramedical staffs from Oman were seconded for experience in Glasgow hospitals. A Senior Radiographer was appointed in October 1984, and a Director of Radiological Services in 1985. A Senior Registrar in Anaesthetics was also appointed to work for a year in Intensive Care Therapy from April 1985, and a Director of Surgical Services was appointed by the Health Ministry in April 1985. Furthermore, a number of nurses were seconded to work in Oman following a request from the Health Minister to the Chief Area Nursing Officer of the Greater Glasgow Health Board.

The College was involved in the planning of all this activity, and met regularly with representatives of the Health Board and the Postgraduate Dean. The College was also involved in establishing patterns of postgraduate medical education based on the export of the successful formula that the College had developed in Glasgow from the 1950s. T. J. Thomson was the College's representative on the Omani Postgraduate Medical Education Committee. This committee staged seminars and clinical meetings in hospitals around Muscat, and in Salalah in the south. Symposia were organised with participants coming from all over the country. For example, in October 1984 a two-day scientific meeting on Gastroenterology was organised jointly by the Health Ministry and the College, with speakers from Glasgow and Kuwait, as well as Oman. The committee also produced the *Oman Medical Newsletter*, which, it was hoped, would form the basis for a national medical journal, as well as advertising the programme of postgraduate activities. The first editor was Dr Abdullah M. Al Riyami, a Glasgow Fellow, and the official College adviser in Oman. In September 1984 Dr Al Khaduri came to Glasgow for his admission to the Fellowship, reemphasising the symbiotic relationship between his country and the College.

During 1985 the overseas sponsorship was revised, and the new scheme further benefited Oman by allowing Omani trainee doctors to be selected for training in the west of Scotland. Under this new double sponsorship scheme, individuals were sponsored by their home authorities to train in proper NHS service posts which were not currently funded due to financial restrictions. The second element was the College, which inspected the posts and supervised the training.

The College also engaged in inspecting posts and recognising where training standards were comparable with the UK, for instance in Abu Dhabi and in Kuwait. This was most useful in identifying suitable standards on which the further organisation of training posts in these countries could be based. Following the approval of certain hospitals, the College was able to hold the first overseas FRCS Part 2 (Final) in Abu Dhabi and Kuwait in 1986. Also in this year the Joint UK Medical Colleges held the first MRCP Part 2 in Hong Kong, with the majority of the input coming from the two Scottish Medical Colleges. In 1987 the written part of the MRCP (UK) Part 2 was held for the first time in Malaysia, and the whole of Part 2 was held in Malaysia in October 1995.

Overseas links were further strengthened when in August 1995 the Presidents of the three royal medical colleges went to Singapore to sign a memorandum of understanding with the National University of Singapore to hold the whole of the Membership Examination there from Spring 1997. In April 1996 the Presidents signed a similar memorandum of understanding with the University of Malaya and the Academy of Medicine of Malaysia to hold the Part 2 of the MRCP (UK) in Malaysia. Most recently, in 1998, the MRCP (UK) Part 2 clinical examination was held in Kuwait for the first time, and the first such examination in Oman took place in October 1998. Also during 1998 Part A of the FRCS was held in Dharwad, India; and Parts A and B were held in the United Arab Emirates.

As the College celebrates its 400th anniversary during 1999 it can take satisfaction from the solidity of its finances and of its recently renovated buildings. But, as both the current President, Mr Colin MacKay, and his immediate predecessor, Professor Norman MacKay, have recently commented, the College, 'is not a building in St Vincent Street. The College is a Fellowship, and we have Fellows all over the world'.[52]

The period covered by this volume of the College history witnessed a dramatic reorientation of the Royal College's role in line with national developments in medical education, and a correspondingly dramatic upturn in its fortunes. The words put into the mouth of the returning shade of key

[52] Interviews by Andrew Hull with Professor Norman MacKay, 24 August 1998, and with Mr Colin McKay, PRCPSG, 11 September 1998.

founder Peter Lowe by the doctor-poet John F. Fergus now ring much more true than they did when they were written in 1918:

> If Maister Peter Lowe were here
> Revisiting this earthly sphere
> What wondrous changes would he see
> Within his famous Faculty ...
> 'Sir, it hath given great joy to me
> To see my infant Faculty
> Grown to so good and great estate.
> The Fellows I congratulate,
> And beg my parting compliments
> To you and future Presidents'.[53]

---

[53] 'Maister Peter Redivivus', verses read at a Faculty Dinner in December 1918, by Dr Freeland Fergus, then President of the RFPSG, in Dr John F. Fergus, *Fancies of a Physician* (Glasgow, 1938), pp. 49–53.

APPENDIX 1

Chronology

1850   The 'Act for Better Regulating the Privileges of the Faculty of Physicians and Surgeons of Glasgow, and amending their Charter of Incorporation' or Faculty Act. Members to be henceforth styled Fellows. Exclusive rights to local jurisdiction in the city of Glasgow and the four counties (Lanark, Renfrew, Dumbarton and Ayr) surrendered in return for reciprocal validity for all surgical licences and diplomas throughout UK and Empire. Faculty establishes a Council as its executive body.

1858   The Medical Act: establishes GMC to oversee all medical examinations which qualify doctor to be included in the new Medical Register. The beginning of the modern profession.

1859   Double Qualification in Medicine and Surgery established with RCPEd.

1862   Faculty moves to Third Faculty Hall at 242 St Vincent Street.

1863   New Faculty Arms containing new motto *non vivere sed valere vita* agreed.

1878   Dentists Act establishes GDC.

1879   Faculty establishes Dental Board to conduct new examination: Licence in Dental Surgery (LDS).

1884   Triple Qualification in Medicine and Surgery established with RCPEd and RCSEd.

1885   An Examination is (re) introduced for the Fellowship.

1886   Medical Act Amendment Act sanctions regional joint examination boards (as for the TQ).

1888   First women to pass TQ.

1892   Reconstruction of 242 St Vincent Street (Faculty Hall).

1902   Purchase of 236 St Vincent Street to extend Faculty Hall.

1909   Faculty becomes Royal Faculty.

1910   Presentation of Faculty Mace by J. Walker Downie.

1911   National Health Insurance Act.

1912   Admission of first female Fellow.

1919   Ministry of Health and Scottish Board of Health established.

1920   Higher Dental Diploma established.

1929   Local Government (Scotland) Act: transfers responsibility for various health matters (including control of Municipal Hospitals) to local town and county councils.

1936   Sir Hector Hetherington becomes Principal and Vice-Chancellor of the University of Glasgow. Immediately begins campaign to modernise Glasgow medicine with full-time professorial appointments in the clinical subjects.

1941   Roy Frew Young's Memorandum to Glasgow Hospitals: Fellowship to be required in consultant appointments.

1944   The Goodenough Report.

1947   George Wishart (first full-time Director of Postgraduate Education) negotiates financial lifeline of £1000 p.a. from University to Faculty to support postgraduate education. Grant continues into the late 1960s.

1948   National Health Service established. University absorbs extra-mural colleges. University Degree becomes *de facto* single portal of first entry to the profession. University also absorbs Dental School and LDS is superseded by BDS.

1949   350th Anniversary of the foundation of the Faculty.

        Hetherington awarded Honorary Fellowship. Diploma in Dental Orthopaedics established.

1950   Medical Act: introduces pre-registration year (comes into force 1953).

1951   Reciprocity of Primary Surgical Fellowship Examinations agreed with RCSEng, RCSEd, and RCSI.

1959   Opening of Maurice Bloch Lecture Theatre.

1960   Glasgow Postgraduate Medical Education Board established. Joint Faculty/University representation, but no longer a University committee.

1961   Nuffield Christ Church (Oxford) Conference urges better organisation of postgraduate training. Clinical tutors appointed to Glasgow hospitals (later called postgraduate advisers).

1962   Provisional Order changes Royal Faculty's name to Royal College. Also amends constitution giving full voting rights to all Fellows (surgery) and all newly styled Members (medicine). Fellowship qua Physician now by election of suitable Members.

1963   Hugh Fraser Reading Room opened.

1967   Fellowship in Dental Surgery supersedes HDD. Dental Council formed.

1968   The Todd Report: expresses consensus that postgraduate training should consist of General Professional followed by Specialist training. Common Membership Part 1 established with MCQs.

1969   MRCP (UK) instituted. Common Part 1 and reciprocity between Part 2s. Joint Committee on Higher Surgical Training established.

1970   Joint Committee on Higher Medical Training established.

1972   Common Part 2 for MRCP (UK) agreed.

1975   Purchase of 234 St Vincent Street further expands College Hall.

1979   College extension officially opened by HRH Princess Alexandra.

1982   Membership in General Dental Surgery established.

1986   Intercollegiate Specialty Examination begins (in Plastic Surgery at first).

1989   Membership in Clinical Community Dentistry established.

Purchase of 232 St Vincent Street.

1990   Dental Fellows become Dental Faculty within RCPSG and Dental Convenor becomes Dean of new Faculty.

1993   The Calman Report on equalising European/UK specialist training times.

1997   Specialist Medical Order: establishes STA; GMC to hold Specialist Register of doctors with CCSTs.

1998   Fiftieth Anniversary of the NHS. Bristol case causes renewed public focus on medical (and especially surgical) qualifications. Colleges to introduce reassessment for hospital doctors? Relative roles of State and Professional bodies in regulation of the medical profession once again debated ...

1999   College celebrates 400 year anniversary.

APPENDIX 2

List of Office-Bearers, 1858–1999

Presidents

1857–1858	Dr James Watson
1858–1859	Dr James Watson
1859–1860	Dr James Watson
1860–1861	Dr William Lyon
1861–1862	Dr William Lyon
1862–1863	Dr Charles Ritchie
1863–1864	Dr Charles Ritchie
1864–1865	Dr Charles Ritchie
1865–1866	Dr John G. Fleming
1866–1867	Dr John G. Fleming
1867–1868	Dr John G. Fleming
1868–1869	Dr Andrew Anderson
1869–1870	Dr Andrew Anderson
1870–1871	Dr John G. Fleming
1871–1872	Dr John G. Fleming
1872–1873	Dr Ebenezer Watson
1873–1874	Dr Ebenezer Watson
1874–1875	Dr Andrew Fergus
1875–1876	Dr Andrew Fergus
1876–1877	Dr Andrew Fergus
1877–1878	Dr Andrew Buchanan
1878–1879	Dr Andrew Buchanan
1879–1880	Dr Andrew Buchanan
1880–1881	Dr R. Scott Orr
1881–1882	Dr R. Scott Orr
1882–1883	Dr R. Scott Orr
1883–1884	Dr Andrew Fergus
1884–1885	Dr Andrew Fergus
1885–1886	Dr Andrew Fergus
1886–1887	Dr James Morton
1887–1888	Dr James Morton
1888–1889	Dr James Morton
1889–1890	Dr Robert Perry
1890–1891	Dr Robert Perry

1891–1892	Dr David Yellowlees
1892–1893	Dr David Yellowlees
1893–1894	Dr David Yellowlees
1894–1895	Dr Bruce Goff
1895–1896	Dr Bruce Goff
1896–1897	Dr Bruce Goff
1897–1898	Sir Hector C. Cameron
1898–1899	Sir Hector C. Cameron
1899–1900	Sir Hector C. Cameron
1900–1901	Dr James Finlayson
1901–1902	Dr James Finlayson
1902–1903	Dr James Finlayson
1903–1904	Mr Henry E. Clark
1904–1905	Mr Henry E. Clark
1905–1906	Dr William L. Reid
1906–1907	Dr William L. Reid
1907–1908	Prof. John Glaister
1908–1909	Prof. John Glaister
1909–1910	Mr David N. Knox
1910–1911	Mr David N. Knox
1911–1912	Dr James A. Adams
1912–1913	Dr James A. Adams
1913–1914	Dr John Barlow
1914–1915	Dr John Barlow
1915–1916	Dr Ebenezer Duncan
1916–1917	Dr Ebenezer Duncan
1917–1918	Dr Ebenezer Duncan
1918–1919	Dr A. Freeland Fergus
1919–1920	Dr A. Freeland Fergus
1920–1921	Dr A. Freeland Fergus
1921–1922	Dr William G. Dun
1922–1923	Dr William G. Dun
1923–1924	Prof. Thomas K. Monro
1924–1925	Prof. Thomas K. Monro
1925–1926	Dr Robert M. Buchanan
1926–1927	Dr Robert M. Buchanan
1927–1928	Mr George H. Edington
1928–1929	Mr George H. Edington
1929–1930	Dr John F. Fergus
1930–1931	Dr John F. Fergus
1931–1932	Mr R. Barclay Ness
1932–1933	Mr R. Barclay Ness
1933–1934	Prof. John M. Munro Kerr
1934–1935	Prof. John M. Munro Kerr

1935–1936	Prof. Archibald Young
1936–1937	Prof. Archibald Young
1937–1938	Dr John Henderson
1938–1939	Dr John Henderson
1939–1940	Dr J. Souttar McKendrick
1940–1941	Mr Roy F. Young
1941–1942	Mr Roy F. Young
1942–1943	Mr James H. MacDonald
1943–1944	Mr James H. MacDonald
1944–1945	Mr William A. Sewell
1945–1946	Mr William A. Sewell
1946–1947	Prof. Geoffrey B. Fleming
1947–1948	Prof. Geoffrey B. Fleming
1948–1949	Dr William R. Snodgrass
1949–1950	Dr William R. Snodgrass
1950–1951	Mr Walter W. Galbraith
1951–1952	Mr Walter W. Galbraith
1952–1953	Mr Andrew J. Allison
1953–1954	Mr Andrew J. Allison
1954–1955	Dr Stanley G. Graham
1955–1956	Dr Stanley G. Graham
1956–1957	Prof. Stanley Alstead
1957–1958	Prof. Stanley Alstead
1958–1959	Mr Arthur H. Jacobs
1959–1960	Mr Arthur H. Jacobs
1960–1961	Dr Joseph H. Wright
1961–1962	Dr Joseph H. Wright
1962–1963	Sir Charles F. W. Illingworth
1963–1964	Sir Charles F. W. Illingworth
1964–1965	Mr Archibald B. Kerr
1965–1966	Mr Archibald B. Kerr
1966–1967	Prof. James H. Hutchison
1967–1968	Prof. James H. Hutchison
1968–1969	Mr Robert B. Wright
1969–1970	Mr Robert B. Wright
1970–1971	Prof. Edward M. McGirr
1971–1972	Prof. Edward M. McGirr
1972–1973	Sir Andrew Watt Kay
1973–1974	Sir Andrew Watt Kay
1974–1975	Sir W. Ferguson Anderson
1975–1976	Sir W. Ferguson Anderson
1976–1977	Prof. Thomas Gibson
1977–1978	Prof. Thomas Gibson
1978–1979	Dr Gavin B. Shaw

1979–1980	Dr Gavin B. Shaw
1980–1981	Mr Douglas H. Clark
1981–1982	Mr Douglas H. Clark
1982–1983	Dr Thomas J. Thomson
1983–1984	Dr Thomas J. Thomson
1984–1985	Mr Ian A. McGregor
1985–1986	Mr Ian A. McGregor
1986–1987	Prof. Arthur C. Kennedy
1987–1988	Prof. Arthur C. Kennedy
1988–1989	Mr James McArthur
1989–1990	Mr James McArthur
1990–1991	Dr Robert Hume
1991–1992	Dr Robert Hume
1992–1993	Sir Donald Campbell
1993–1994	Sir Donald Campbell
1994–1995	Prof. Norman MacKay
1995–1996	Prof. Norman MacKay
1996–1997	Prof. Norman MacKay
1997–1998	Mr Colin MacKay
1998–1999	Mr Colin MacKay

Visitors

1857–1858	Dr William Lyon
1858–1859	Dr William Lyon
1859–1860	Mr George Watt
1860–1861	Mr George Watt
1861–1862	Mr George Watt
1862–1863	Dr John G. Fleming
1863–1864	Dr John G. Fleming
1864–1865	Dr John G. Fleming
1865–1866	Dr Andrew Anderson
1866–1867	Dr Andrew Anderson
1867–1868	Dr Andrew Anderson
1868–1869	Dr Harry Rainy
1869–1870	Dr Harry Rainy
1870–1871	Dr William Weir
1871–1872	Dr George Buchanan
1872–1873	Dr George Buchanan
1873–1874	Dr Andrew Fergus
1874–1875	Dr James Morton
1875–1876	Dr James Morton
1876–1877	Dr John B. Cowan
1877–1878	Dr R. Scott Orr
1878–1879	Dr R. Scott Orr
1879–1880	Dr R. Scott Orr
1880–1881	Dr Robert Perry
1881–1882	Dr Robert Perry
1882–1883	Dr Robert Perry
1883–1884	Dr Henry Muirhead
1884–1885	Dr Henry Muirhead
1885–1886	Dr Henry Muirhead
1886–1887	Dr Alexander Robertson
1887–1888	Dr Alexander Robertson
1888–1889	Dr Alexander Robertson
1889–1890	Dr David Yellowlees
1890–1891	Dr David Yellowlees
1891–1892	Dr Bruce Goff
1892–1893	Dr Bruce Goff
1893–1894	Dr Bruce Goff
1894–1895	Dr James Dunlop
1895–1896	Dr James Dunlop
1896–1897	Dr Hector C. Cameron
1897–1898	Dr Adam L. Kelly
1898–1899	Dr Adam L. Kelly

1899–1900	Dr James Finlayson
1900–1901	Dr John Burns
1901–1902	Dr John Burns
1902–1903	Mr Henry E. Clark
1903–1904	Dr William L. Reid
1904–1905	Dr William L. Reid
1905–1906	Dr John Glaister
1906–1907	Dr John Glaister
1907–1908	Mr David N. Knox
1908–1909	Mr David N. Knox
1909–1910	Dr James A. Adams
1910–1911	Dr James A. Adams
1911–1912	Dr John Barlow
1912–1913	Dr John Barlow
1913–1914	Dr Ebenezer Duncan
1914–1915	Dr Ebenezer Duncan
1915–1916	Mr J. Walker Downie
1916–1917	Dr A. Freeland Fergus
1917–1918	Dr A. Freeland Fergus
1918–1919	Dr William G. Dun
1919–1920	Dr William G. Dun
1920–1921	Dr William G. Dun
1921–1922	Dr William G. Dun
1922–1923	Dr William R. Jack
1923–1924	Mr James H. Pringle
1924–1925	Mr James H. Pringle
1925–1926	Dr George H. Edington
1926–1927	Dr George H. Edington
1927–1928	Dr John F. Fergus
1928–1929	Dr John F. Fergus
1929–1930	Mr R. Barclay Ness
1930–1931	Mr R. Barclay Ness
1931–1932	Dr John M. Munro Kerr
1932–1933	Dr John M. Munro Kerr
1933–1934	Mr Archibald Young
1934–1935	Mr Archibald Young
1935–1936	Dr John Henderson
1936–1937	Dr John Henderson
1937–1938	Dr J. Souttar McKendrick
1938–1939	Dr J. Souttar McKendrick
1939–1940	Mr Roy F. Young
1940–1941	Mr James H. MacDoanld
1941–1942	Mr James H. MacDonald
1942–1943	Mr William A. Sewell

1943–1944	Mr William A. Sewell
1944–1945	Dr Geoffrey B. Fleming
1945–1946	Dr Geoffrey B. Fleming
1946–1947	Mr J. Scouler Buchanan
1947–1948	Dr William R. Snodgrass
1948–1949	Mr Walter W. Galbraith
1949–1950	Mr Walter W. Galbraith
1950–1951	Mr Andrew J. Allison
1951–1952	Mr Andrew J. Allison
1952–1953	Dr Stanley G. Graham
1953–1954	Dr Stanley G. Graham
1954–1955	Prof. Stanley Alstead
1955–1956	Prof. Stanley Alstead
1956–1957	Mr Arthur H. Jacobs
1957–1958	Mr Arthur H. Jacobs
1958–1959	Dr Joseph H. Wright
1959–1960	Dr Joseph H. Wright
1960–1961	Sir Charles F. W. Illingworth
1961–1962	Sir Charles F. W. Illingworth
1962–1963	Mr Archibald B. Kerr
1963–1964	Mr Archibald B. Kerr
1964–1965	Prof. James H. Hutchison
1965–1966	Prof. James H. Hutchison
1966–1967	Mr Robert B. Wright
1967–1968	Mr Robert B. Wright
1968–1969	Prof. Edward M. McGirr
1969–1970	Prof. Edward M. McGirr
1970–1971	Sir Andrew Watt Kay
1971–1972	Sir Andrew Watt Kay
1972–1973	Sir Andrew Watt Kay
1973–1974	Sir W. Ferguson Anderson
1974–1975	Prof. Thomas Gibson
1975–1976	Prof. Thomas Gibson
1976–1977	Dr Gavin B. Shaw
1977–1978	Dr Gavin B. Shaw
1978–1979	Mr Douglas H. Clark
1979–1980	Mr Douglas H. Clark
1980–1981	Dr Thomas J. Thomson
1981–1982	Dr Thomas J. Thomson
1982–1983	Mr Ian A. McGregor
1983–1984	Mr Ian A. McGregor
1984–1985	Prof. Arthur C. Kennedy
1985–1986	Prof. Arthur C. Kennedy
1986–1987	Mr James McArthur

1987–1988	Mr James McArthur
1988–1989	Sir Abraham Goldberg
1989–1990	Dr Robert Hume
1990–1991	Sir Donald Campbell
1991–1992	Sir Donald Campbell
1992–1993	Prof. Norman MacKay
1993–1994	Prof. Norman MacKay
1994–1995	None
1995–1996	None
1996–1997	Mr Colin MacKay
1997–1998	None
1998–1999	None

Honorary Secretaries

1954–1958	Dr Alexander H. Imrie
1958–1965	Dr Gavin B. Shaw
1965–1973	Dr Thomas J. Thomson
1973–1983	Dr Norman MacKay
1983–1991	Dr Alistair D. Beattie
1991–1995	Dr Brian O. Williams
1995–1998	Dr Stefan D. Slater
1998–1999	Dr Colin G. Semple

Honorary Treasurers

1857–1860	Mr George Watt
1860–1865	Dr William Weir
1865–1879	Dr John Coats
1879–1901	Dr J. D. MacLaren
1901–1904	Dr William L. Reid
1904–1918	Dr William G. Dun
1918–1922	Mr George McIntyre
1922–1939	Mr James H. MacDonald
1939–1944	Mr William J. Richard
1944–1948	Mr Walter W. Galbraith
1948–1955	Mr Matthew White
1955–1959	Dr James A. W. McCluskie
1959–1966	Mr Robert B. Wright
1966–1970	Mr Kenneth Fraser
1970–1976	Mr Alistair J. Mack
1976–1986	Mr Colin MacKay
1986–1993	Mr Douglas G. Gilmour
1993–1999	Mr Ian G. Finlay

Honorary Librarians

1858–1860	Dr James Adams
1860–1861	Dr James McGhie
1861–1863	Dr William Brown
1863–1865	Dr Andrew B. Buchanan
1865–1869	Dr George Rainy
1869–1875	Dr James D. MacLaren
1875–1901	Dr James Finlayson
1901–1909	Dr J. Lindsay Steven
1909–1919	Dr Alexander Napier

1919–1933	Dr Ebenezer H. L. Oliphant
1933–1946	Dr William R. Snodgrass
1946–1963	Dr Archibald L. Goodall
1963–1974	Prof. Thomas Gibson
1974–1980	Sir Charles F. W. Illingworth
1980–1994	Dr Roderick N. M. MacSween
1994–1999	Prof. Ian T. Boyle

Bibliography

Listed here are the most important sources for Glasgow, Glasgow medicine and the RCPSG in the period covered by this volume

Wendy Alexander, *First Ladies of Medicine: The Origins, Education and Destination of Early Women Medical Graduates of Glasgow University* (Glasgow, 1987).

Stanley Alstead, *RCPSG,: Extracts from the Minute Books of the Faculty/College from 1798–1958 with Comments by the Compiler*, 3 vols, RCPSG, 1/13/3/1.

R. D. Anderson, 'Scottish University Professors, 1880–1939: Profile of an Elite', *Scottish Economic and Social History*, 7 (1987), pp. 27–54.

J. L. Brand, *Doctors and the State* (Baltimore, 1965).

John Butt, *John Anderson's Legacy: The University of Strathclyde and its Antecedents, 1796–1996* (East Linton, 1996).

J. D. Comrie, *History of Scottish Medicine*, 2 vols (London, 1932).

H. Conway and R. T. Hutcheson, 'The Glasgow Medical Faculty, 1936–39: A Change of Direction', *Scottish Medical Journal*, 41 (1996), pp. 178–79.

A. M. Cooke, *A History of the Royal College of Physicians of London*, iii (Oxford, 1972).

Zachary Cope, *The Royal College of Surgeons of England* (London, 1959).

James Coutts, *A History of the University of Glasgow, 1451–1909* (Glasgow, 1909).

W. S. Craig, *History of the Royal College of Physicians of Edinburgh* (Oxford, 1976).

C. H. Cresswell, *The Royal College of Surgeons of Edinburgh: Historical Notes from 1505–1905* (Edinburgh, 1926).

Anne Crowther and Brenda White, *On Soul and Conscience: The Medical Expert and Crime. 150 Years of Forensic Medicine in Glasgow* (Aberdeen, 1988).

Anne Digby, *Making a Medical Living* (Cambridge, 1994).

Derek Dow and Michael Moss, 'The Medical Curriculum at Glasgow in the Early Nineteenth Century', *History of Universities*, 7 (1988), pp. 227–57.

Derek Dow and Kenneth Calman (eds), *The Royal Medico-Chirurgical Society of Glasgow: A History, 1814–1989* (Glasgow, 1989).

Robin Dowie, *Postgraduate Medical Education and Training: The System in England and Wales* (London, 1987).

Alexander Duncan, *Memorials of the Faculty of Physicians and Surgeons of Glasgow, 1599–1850* (Glasgow, 1896).

P. R. Fleming, W. G. Manderson, M. B. Matthews, P. H. Sanderson and J. F. Stokes, 'Evolution of an Examination: MRCP (UK)', *BMJ*, 13 April 1974, pp. 99–107.

Michael Foot, *Aneurin Bevan*, ii (London, 1973).

Daniel Fox, *Health Policies, Health Politics: The British and American Experience, 1911–1965* (Princeton, 1986).

W. Hamish Fraser and Irene Mavor (eds), *Glasgow, ii, 1830–1912* (Manchester, 1996).

Robert Campbell Garry (edited by David Smith), *Life in Physiology: Memoirs of Glasgow University's Institute of Physiology during the 1920s and 1930s* (Glasgow, 1992).

Tom Gibson, *The Royal College of Physicians and Surgeons of Glasgow: A Short History Based on the Portraits and Other Memorabilia* (Edinburgh, 1983).

David Hamilton, *The Healers: A History of Medicine in Scotland* (Edinburgh, 1981).

S. W. F. Holloway, 'Medical Education in England, 1830–1858: A Sociological Analysis', *History*, 49 (1964), pp. 299–324.

Frank Honigsbaum, *The Division in British Medicine* (London, 1979).

Frank Honigsbaum, *Health, Happiness and Security: The Creation of the National Health Service* (London, 1989).

Andrew Hull, 'Hector's House: Sir Hector Hetherington and the Academicisation of Glasgow Medicine before the National Health Service', *Medical History* (forthcoming, 2000).

Andrew Hull, audio tapes and transcripts of interviews with eminent members of the Glasgow medical profession conducted during 1997–98, RCPSG, 1/13/9.

R. T. Hutcheson, *The University of Glasgow, 1920–1974: The Memoir of Robert T. Hutcheson* (Glasgow 1997).

John Hutchison, *RFPS/RCPS: Property and Environment*, RCPSG, 1/13/7/9.

Sir Charles Illingworth, *University Statesman: Sir Hector Hetherington* (Glasgow, 1971).

Sir Charles Illingworth, *There is a History in All Men's Lives* (Milngavie, 1988).

Stephen Jacyna, 'The Laboratory and the Clinic: The Impact of Pathology on Surgical Diagnosis in the Glasgow Western Infirmary, 1875–1910', *Bulletin for the History of Medicine*, 62 (1988), pp. 384–406.

Jacqueline Jenkinson, *Scottish Medical Societies, 1731–1939: Their History and Records* (Edinburgh, 1993).

Jacqueline Jenkinson, Michael Moss and Iain Russell, *The Royal: The History of the Glasgow Royal Infirmary, 1974–1994* (Glasgow, 1994).

Christopher Lawrence, 'Incommunicable Knowledge: Science, Technology and the Clinical Art in Britain, 1850–1914', *Journal of Contemporary History*, 20 (1985), pp. 503–20.

Irvine Loudon, *Medical Care and the General Practitioner, 1750–1850* (Oxford, 1986).

Irvine Loudon, 'Medical Practitioners 1750–1850 and the Period of Medical Reform in Britain', in Andrew Wear (ed.), *Medicine in Society* (Cambridge, 1992), pp. 219–47.

Irvine Loudon 'Medical Education and Medical Reform', in V. Nutton and R. Porter (eds), *The History of Medical Education in Britain* (Amsterdam, 1995).

Edith MacAlister, *Sir Donald MacAlister of Tarbert* (London, 1935).

F. A. Macdonald, 'Vaccination Policy of the Faculty of Physicians and Surgeons of Glasgow, 1801 to 1863', *Medical History*, 44 (1997), pp. 291–321.

Alexander MacGregor, *Public Health in Glasgow, 1905–1946* (Edinburgh, 1967).

Peter Mackenzie, transcripts of interviews with eminent members of the Glasgow medical profession, RCPSG, 18.

J. D. Mackie, *The University of Glasgow, 1451–1951* (Glasgow, 1954).

Gordon McLachlan (ed.), *Improving the Common Weal: Aspects of Scottish Health Services, 1900–1984* (Edinburgh, 1987).

Loudon MacQueen and Archibald B. Kerr, *The Western Infirmary, 1874–1974: A Century of Service to Glasgow* (Glasgow, 1974).

Ministry of Health, *Report of the Inter-Departmental Committee on Medical Schools* (the Goodenough Report) (London, 1944).

Charles Newman, *The Evolution of Medical Education in the Nineteenth Century* (London, 1957).

N. and J. Parry, *The Rise of the Medical Profession* (London, 1976).

F. N. L. Poynter (ed.), *The Evolution of Medical Education in Britain* (London, 1966).

Royal Commission on Medical Education (the Todd Report), Cmnd 3569 (London, 1968).

Gavin Shaw, 'The Royal College of Physicians and Surgeons of Glasgow', in Dow and Calman (eds), *The Royal Medical-Chirurgical Society of Glasgow: A History, 1814–1989* (Glasgow, 1989), pp. 102–6.

David Smith and Malcolm Nicolson, 'The "Glasgow School" of Paton, Findlay and Cathcart: Conservative Thought in Chemical Physiology, Nutrition and Public Health', *Social Studies of Science*, 19 (1989), pp. 195–238.

David Smith and Malcolm Nicolson, 'Chemical Physiology versus Biochemistry, the Clinic versus the Laboratory: The Glaswegian Opposition to Edward Mellanby's Theory of Rickets', *Proceedings of the Royal College of Physicians of Edinburgh*, 19, (January 1989), pp. 51–60.

Margaret Stacey, *Regulating British Medicine: The General Medical Council* (Chichester, 1992).

Rosemary Stevens, *Medical Practice in Modern England: The Impact of Specialization and State Medicine* (New Haven, 1966).

T. N. Stokes, 'A Coleridgean against the Medical Profession: John Simon and the Parliamentary Campaign for the Reform of the Medical Profession, 1854–8', *Medical History*, 33 (1989), pp. 343–59.

Ivan Waddington, 'General Practitioners and Consultants in Early Nineteenth Century England: The Sociology of an Intra-Professional Conflict', pp. 164–88, in J. Woodward and D. Richards (eds), *Health Care and Popular Medicine in Nineteenth Century England* (London, 1977).

Ivan Waddington, *The Medical Profession in the Industrial Revolution* (Dublin, 1984).

A. Lockhart Walker, *The Revival of the Democratic Intellect: Scotland's University Traditions and the Crisis in Modern Thought* (Edinburgh, 1994).

Oliver M. Watt, *Stobhill Hospital: The First Seventy Years* (Glasgow, 1971).

Charles Webster, *The Health Services since the War*, i, *Problems of Health Care: The National Health Service before 1957* (London, 1988).

Charles Webster, *The Health Services since the War*, ii, *Government and Health Care: The National Health Service, 1958–1979* (London, 1996).

Index

Illustrations are shown in bold

MacKay, Prof. Norman 215, 233–34, 236, 238, 247, 250, 252
MacKelvie, Dr A. A. 185n.
McKendrick, Dr John Souttar 100, 246, 249
Mackenzie, Dr A. 185n.
Mackenzie, Prof. William 60 and n.; *A Treatise on Diseases of the Eye* 60n.; *The Physiology of Vision* 60n.
Mackey, Prof. William Arthur 95–98, 96n., 135n., 147 and n., 181n.
Mackie, Dr E. Gordon 85
MacKinnon, Grace C. C. 46
Mackintosh, Charles Rennie 58n.
MacLaren, Dr James D. 252
Maclean, Dr Ewen 71
MacLeod, Prof. George H. B. 28
Macnee, Daniel, painter: pl. 5
McNee, Sir John William 106, 115, 126, 136n., 203
MacSween, Dr Roderick N. M. 253
McVail, John C. 43
Madras 234
Malaysia 238
Manchester 63, 99, 173
Manderson, W. G. 161n., 203, 214
Marischal College 12 and n.; *see also* Aberdeen University
Martial (M. V. Martialis), *Epigrams* 4n.
Master of Surgery (Chirurgiae Magister) degree 18–19, 51
Maternity and Infant Welfare Officers 81
Maurice Bloch Lecture Theatre: xxx, 158–59
MB (Bachelor of Medicine) degree 18, 19
MCQ — multiple choice questions 196, 198–200, 203, 204, 206, 226, 230
MD (Doctor of Medicine) degree 18, 51
Mearnskirk Hospital, Glasgow 147, 160, 161n.
MEC, *see* Medical Education Committees
Medical Act (1858): John Simon and 5; unresolved issues 5, 6; confusion caused by 8, 10–11, 12–16, 17–18, 19; amendment Bills 20–23, 36–39; 1879–80 select committee 26, 38 and n.; 1881–82 Royal Commission 1, 38, 41, 47, 99; 1886 Amendment Act 6, 47; effects on Faculty: xv, 49; GMC powers under 31; Schedule A licensing bodies 2, 4, 17, 36,

42; undergraduates and 120, 209; 1876 sex discrimination Act 46; 1878 Dentists Act 47; *see also* General Medical Council
Medical Act (1950) 120
Medical Act (1978) 120, 224–25
Medical Act Amendment Act (1886) 6, 47
Medical Advisory Committee (MAC) 126, 127, 195, 203–4, 234
Medical Advisory Committee on Silicosis 75
Medical Alliance Association 40
Medical Appeals Tribunals 75
medical conferences: Oxford 182–85; Melbourne 198; Oman 237; symposia 146–49, 177, 189, 234
medical corporations, *see* Royal Colleges
medical degrees: xv, xvi, 17–19, 47, 51, 55, 63, 77, 201; *see also* qualifications; universities
medical discipline 58, 224–25, 228–29
Medical Education Committees (MEC) 89–90, 90n., 106, 116, 130; *see also* Regional Hospital Boards; undergraduate education; Western Regional Hospital Board
medical ethics, *see* medical discipline; unethical practice
medical examinations (to qualify), *see* curricula and examinations
Medical Officers of Health (MOH) 11n., 52 and n., 53, 61n., 77; *see also* public health
Medical Poor Law Officers 12; *see also* Poor Law Board
Medical Reform Committee 40–41; *see also* British Medical Association
Medical Register 10, 20–21, 38, 46, 47; *see also* General Medical Council; qualifications
medical schools, *see* Anderson's; St Mungo's; teaching hospitals
Medical Staff Associations 188
Medical Statistics of Life Assurance, The (Fleming) 3n.
Medical Teachers Association 20
'Medicine Today' BBC programme 190
Medico-Chirurgical Society (RMCSG) 77, 95n., 104, 136, 137, 144, 148
Medico-Psychological Association 77, 195–96

113n. –114n.; *see also* women
practitioners
Shaw, Dr Gavin Brown: pl. 11; background
135–36n., 147n.; Presidency 246–47;
Visitor 250; Hon. Secretary 158, 200,
252; progressive reformer 135–36, 152,
158; influence of younger Fellows 142,
143n., 150, 160, 174; subscriptions
committee 136; postgraduate medical
education 142–44, 144n., 159, 175, 186,
223n.; medical conference organisation
146–48, 189; Ward Round Committee
160–61, 160n., 161n.; 'Recent Advances'
courses 161–62; television work 189,
190; the GPGMEC and 162, 176–77,
178–79, 180; views on Oxford
conference 185, 186; training reform
174, 175, 197; clinical tutor 185n.;
Membership debate 200–201, 202,
212–13; Todd Report 212–13; RFPSG
name-change 192–93; personal papers
XVIII
Shaw, George Bernard, *The Doctor's
Dilemma* 57
Shearer (Gilbert) and Co.: XXVI
Sheffield University 63
SHMO — Senior Hospital Medical Officer
grade 155, 156; *see also* hospital
appointments
SHO — Senior House Officers 161, 162,
183; *see also* hospital appointments
Short, Dr I. A. 185n.
Silicosis, Medical Advisory Committee on 75
Sillar, William 160
Silvercraigs, Saltmarket: XXII
Simmons, Williemina 47
Simon, Sir John: background 20 and n., 37;
1858 Medical Act 5; quest for a single
portal: xv, xvi, 2, 5–8, 22, 37; General
Medical Council and 6–7, 20, 21, 22, 23;
renewed reform agitation 20–22, 23, 31,
37; RCSEd Council 20
Singapore 235, 238
single licence: Diploma in Surgery 14;
curriculum 14, 15, 17, 41–42, 44;
'half-qualification' 20–21, 22, 39–40;
GMC inspected examinations 31, 32–34,
35–36; in defence of 41–42, 41n.; fees 15,

17, 32; overseas students and 27;
superseded by Triple Qualification 42,
43; *see also* curricula and examinations;
Double Qualification; Fellowship;
Membership; qualifications; Triple
Qualification
single portal of entry, *see* qualifications
Sir Maurice Bloch Trust 159
'Six Cardinal Points' 67, 68, 69, 70, 73
SJC, *see* Standing Joint Committee
Slater, Dr Stefan D. 234, 252
slavery: XXIV–XXV, XXVn.
Slessor, Dr A. 185n.
Sloan, A. 63
slums 74–75
smallpox 75–76; *see also* public health;
vaccinations
Smith, David 115
SMO — Specialist Medical Order 112, 217
Snodgrass, Dr William Robertson: pl. 8;
medical background 75n. –76n.;
Presidency 246; Visitor 250; Hon.
Librarian 253; 1920 smallpox outbreak
75–76; Dr A.M.K.Barron and 113, 113n.
–114n.; postgraduate medical education
115; Committee of Physicians 126
Society of Apothecaries 24, 40; *see also*
Apothecaries' Company; general
practitioners
South Africa 134, 192, 200, 217
Southern General Hospital, Glasgow: Poor
Law institution 75n., 113n., 143n.;
Hetherington and 105–6, 184; teaching
hospital 95n., 143n., 160, 223; symposia
sessions 147; Ward Round courses 160
and n.; Institute of Neurological Sciences
173n.
Southern Medical Society 61n.
Spanish influenza pandemic (1918–19) 75
specialists (general), *see* consultants and
specialists
Specialist Advisory Committee on Dental
Surgery and Orthodontics 207
Specialist Advisory Committees (SAC)
214–15
Specialist Medical Order (SMO) 112, 217
Specialist Register 227; *see also* General
Medical Council; qualifications